Building a Professional Teaching Identity on Social Media

Building a Professional Teaching Identity on Social Media

A Digital Constellation of Selves

Janine S. Davis
University of Mary Washington, USA

SENSE PUBLISHERS
ROTTERDAM/BOSTON/TAIPEI

A C.I.P. record for this book is available from the Library of Congress.

ISBN: 978-94-6300-700-9 (paperback)
ISBN: 978-94-6300-701-6 (hardback)
ISBN: 978-94-6300-702-3 (e-book)

Published by: Sense Publishers,
P.O. Box 21858,
3001 AW Rotterdam,
The Netherlands
https://www.sensepublishers.com/

All chapters in this book have undergone peer review.

Printed on acid-free paper

All Rights Reserved © 2016 Sense Publishers

No part of this work may be reproduced, stored in a retrieval system, or transmitted in any form or by any means, electronic, mechanical, photocopying, microfilming, recording or otherwise, without written permission from the Publisher, with the exception of any material supplied specifically for the purpose of being entered and executed on a computer system, for exclusive use by the purchaser of the work.

*Thank you to all who made this book possible. This book is for you.
For the Schanks and the Davises.
For internet friends both close and far-flung.
For people I have never met, not even once.
For all the students I have ever taught,
and who continue to teach me.
For my friends, especially Jasmine, Chrissy, and Vicki.
For Alex, Ayla, and Lena.*

TABLE OF CONTENTS

Foreword 　*Martha Burtis*	xi
List of Tables	xv
List of Figures	xvii
Chapter 1: Why Social Media?	1
Using This Text	3
Chapter 2: Research Support and Conceptual Framework	5
Research Support	5
Conceptual Framework	5
Summary	9
Extension Activities	10
Chapter 3: Social Media: A Comparison	11
Details about Social Media Use	12
Blurring the Lines	15
Challenges	18
Summary	20
Extension Activities	20
Chapter 4: Professional Development Opportunities on Social Media	21
Building a Professional Learning Network	22
Summary	27
Extension Activities	27
Chapter 5: Social Media in the Classroom	29
Calls for Use of Social Media in Education	30
College-Level Uses	31
K12-Level Uses	32
Summary	34
Extension Activities	35
Chapter 6: Developing an Identity on Social Media	37
The Fragmentation of Online Identities	37
Social Media Identities: Not Always a Choice	38

TABLE OF CONTENTS

    Your Varying Identities   39
    Summary   42
    Extension Activities   42

Chapter 7: Developing a Professional Identity on Social Media   45

    Professional Uses   46
    Summary   49
    Extension Activities   50

Chapter 8: Effects of Your Emerging Identity   51

    Effects on Oneself   51
    Effects on Others   53
    Summary   54
    Extension Activities   55

Chapter 9: Methods of the Study   57

    Research Questions   57
    Methods: Data Collection   57
    Methods: Data Analysis   60
    Limitations   61
    Extension Activities   63

Chapter 10: Findings: The Survey   65

    Extension Activities   70

Chapter 11: Case Study One: Nora   71

    Tweets   71
    Nora's Story   71
    Extension Activities   78

Chapter 12: Case Study Two: Marina   79

    Tweets   79
    Marina's Story   79
    Extension Activities   83

Chapter 13: Case Study Three: Callie   85

    Tweets   85
    Callie's Story   85
    Extension Activities   89

TABLE OF CONTENTS

Chapter 14: Case Study Four: Kate 91
    Tweets 91
    Kate's Story 91
    Extension Activities 95

Chapter 15: Cross-Case Analysis and Discussion 97
    A Way of Life 97
    Tensions and Opportunities 99
    Persona and Identity Formation 101
    Summary 104
    Extension Activities 104

Chapter 16: Implications 105
    Implications for Beginning and Practicing Teachers 105
    Implications for Teacher Educators 106
    Implications for Researchers 107
    Summary 108
    Extension Activities 108

Chapter 17: Conclusions and Next Steps 109
    Extension Activities 111

References 113

MARTHA BURTIS

FOREWORD

In 2011, Ashley Payne, a high school English teacher in Georgia, was forced to resign after the parent of one of her students discovered a photo on Payne's Facebook wall of her posing with alcoholic beverages. The picture, taken several years earlier while Payne was traveling, is pretty tame by the standards of what many young adults post on social media. Payne's case made national news, and her legal attempts to be reinstated were unsuccessful. Her story is one in a long string of narratives about educators and social media. A quick Google search will turn up dozens, if not hundreds, of anecdotes involving teachers fired for the content of what they shared on various social media accounts.

Faced with these chilling stories, it's difficult to know what to advise a preservice teaching student, ready to embark on a career in education. It is tempting to suggest that one should proceed with extreme caution, perhaps avoiding social media altogether. However, preservice teachers often are also twenty-something young adults, only a few years out of high school, still forming their identities and determining how to best present themselves in face-to-face and online situations. Moreover, social media is woven into the fabric of their lives and has been since many of them were preteens. This combination of identity formation coupled with the availability and ubiquity of social media is an undeniable reality of our particular moment in higher education. Turning a blind eye to this reality is an abdication of our responsibilities as educators. If you're not sure where to begin as you consider the integration of social media into the education curriculum, this book can serve as a guide.

Returning again to the chilling narratives of teachers being fired, it is easy to be shocked by the reported details. Some of the stories, like Payne's, seem to represent an overblown reaction to a relatively minor transgression. Other stories, such as a teacher in Louisiana who was fired after using her Twitter account to share strong criticisms of Common Core State Standards, could be interpreted as a retaliation against a particular political or professional stance. In contrast to these troubling tales, there are also incidents in which (usually young, or early-career) educators are reported to have posted sexually explicit, racially charged, or simply vulgar status updates, photos, or videos. It's easy to have an initial reaction of shock and outrage. However, anyone who has ever followed the social media accounts of college students might view these incidents with a different eye.

M. BURTIS

In my work with students, I frequently find myself following them on Twitter, Facebook, and other social media and networking sites. This is in large part because of the kinds of classes I teach which deal with digital identity, digital storytelling, and online community. Even knowing that I'm watching, my students make use of vulgar and profane language and provide details of their lives that I would generally classify as "over-sharing." I have to remind myself that these students are at a different moment in their lives than I am. They are usually of traditional college age. At my school, they are also most likely away from home in a residential university setting for the first time. They are encountering people, situations, and particular challenges that are different from anything they've experienced before. Social media serves as a performative space in which they are trying out new identities (often with the trappings of new language, customs, and behaviors). They may try out several aspects of identity before their own sense of themselves begins to solidify. While this experimentation occurs, they may say and share things that are out of "the norm" for them. The flip side to this situation is that in a few short years, I know that they will be graduating and taking on other kinds of personae. They will have to begin, to borrow their vernacular, "adulting" soon.

What is our role—what is our responsibility, in fact—to these students when it comes to helping them navigate the world of social media? And, in the world of preservice teachers, where the stakes may seem even higher, how can we prepare students to be successful? Davis' text points us in important and useful directions. In the classroom, we need to promote thoughtful engagement with the tools of social media. Beyond this, we need to help our students to articulate the different aspects of their identities and to reflect upon how they enact and expose those identities in authentic, deliberate ways. Our students are already capable of understanding that they have different audiences in their lives, and that these audiences need to be addressed in different ways. However, they still need the experience, language, and information to make wise choices about the presentation of their own digital identity.

Davis also takes us beyond the (not insignificant) challenge of helping students to hone and more carefully consider their digital identity. She clearly articulates that avoiding the conversation with our students about social media does them a huge disservice by divorcing them from vibrant, vital online communities of educators. For example, from relatively early in its existence, Twitter has served as an online hub for the development of personal learning networks. Davis offers clear advice and strategies for fostering and developing one's own network, using Twitter as an effective catalyst. By turning the conversation about social media away from the reactionary ("How do I avoid getting in trouble?") to the proactive ("How can these spaces and tools make me a better teacher?") she also recasts these issue for us. Social media isn't something we merely have to overcome; it is something we need to master.

The preservice teachers in Davis' book are both students and teachers in training. As students, they have an opportunity to learn and think more deeply about how

FOREWORD

online and digital spaces are changing the culture we live in, and how they, as student teachers will respond to that culture. As teachers in training, they're also arming themselves with the knowledge and expertise to educate the next generation of college students about how to thoughtfully, critically, and humanely engage in digital spaces. After reading this book, I'm hopeful that our future teachers can thoughtfully and critically navigate social media, and, in doing so, they can provide the students they teach with a bright example of what's possible.

LIST OF TABLES

Table 1. A comparison of social media sites   12
Table 2. Enjoyment and time of use   66

LIST OF FIGURES

Figure 1.	Tweet from Smithsonian Air and Space Museum (Smithsonian, 2013)	1
Figure 2.	Tweet from the Getty Museum (J. Paul, 2016)	1
Figure 3.	Preservice teachers' development of persona and identity (adapted from Davis, 2010)	7
Figure 4.	Uses and purposes of social media as based on Tufekci (2008) and Wright et al. (2014)	9
Figure 5.	Social networking site use by age group, 2005–2013 (Pew Research Center, 2013, para. 5)	13
Figure 6.	Tweet with emojis @girlposts in response to a retweeted, complimentary tweet (Common, 2016)	14
Figure 7.	Retweet of author Mary Karr's tweet with added hashtags (Davis, 2016)	17
Figure 8.	Example of tweet compiling tool as shared in a tweet to students (Davis, 2013)	23
Figure 9.	Example of meme, "unhelpful high school teacher" (Unhelpful, n.d.)	24
Figure 10.	Sources to follow	25
Figure 11.	Tweet to course hashtags and colleagues (Davis, 2016)	32
Figure 12.	School district tweet with emoji (Wake, 2016)	33
Figure 13.	Personal identity factors	41
Figure 14.	Professional identity interactions	49

CHAPTER 1

WHY SOCIAL MEDIA?

I fell in love with Twitter on Super Bowl Sunday of 2013. Recalling it is easy, because the Getty Museum account (@GettyMuseum), which I had happened to follow when Twitter suggested various accounts based on my interests, started using the #MuseumSuperBowl hashtag to show images of super bowls in their collection. Other museums caught on and did the same, and soon the tweets paralleled the events of the game, such as when the lights went out in the stadium and the Smithsonian Air and Space Museum tweeted this:

> **SmithsonianAirSpace** @airandspace · 3 Feb 2013
> In the dark? Imagine being on Apollo 12 when they lost electrical power during launch: s.si.edu/hnNxh #MuseumSuperBowl
> ↰   ↻ 47   ★ 9   ...

Figure 1. Tweet from Smithsonian Air and Space Museum (Smithsonian, 2013)

Suddenly, it became clear that there was not only a living, breathing human but also one with a sense of humor and fun associated with the accounts of organizations such as the Getty Museum and the Smithsonian Air and Space Museum, and the organizations took on those same endearing qualities (Davis, 2013). More recently the Getty Museum and other museums have interacted with the public using hashtags such as #musEmoji and #MuseumSelfie.

> **J. Paul Getty Museum** @GettyMuseum · Jun 22
> To celebrate #musEmoji day, we're sharing our fav labors of Hercules using ONLY emojis. Guess which labor each emoji-story tells—good luck! 💀
> ↰   ↻ 14   ♥ 22   ...

Figure 2. Tweet from the Getty Museum (J. Paul, 2016)

Enacting various identities on social media is also possible for individuals; it occurs every second of every day. The effect we have on ourselves and others through our posts and comments can be intense and life changing or simple and mundane, but our audiences are an inextricably linked aspect of how we create and display our personae and identities.

How did I find Twitter in the first place? At the University of Mary Washington, we have a program called Domain of One's Own (DoOO), where students, faculty, and staff alike are encouraged to develop an online presence (University of Mary Washington, n.d.). Twitter is just one of the tools that participants in the DoOO program may choose to use as a component of their online web presence; blogs are also common features of DoOO sites (see Davis, 2013).

How and Why We Use Social Media

Every day, millions of people, some close friends and family, and others complete strangers to each other, interact and share information electronically, through social media sites such as Twitter, Facebook, and Instagram. We choose profile pictures, list biographical details, and construct personae that we demonstrate with various avatars and a constant stream of text and images. These public and digital representations of our selves are formed in similar ways to theatrical productions; we act on a kind of stage when we interact on social media (Tu, Blocher, & Roberts, 2008). New sites and applications appear, morph, and disappear with each coming month, but connecting with others on the internet in various ways is a concept that's here to stay. The connections that individuals explore on social media can inform the work of teachers as they plan and deliver lessons and as they form professional identities.

Love is an apt metaphor for one's interaction with social media because so much depends upon factors within and outside one's control. Are you interacting with a narrow circle of people? You may never like or love any one of them; you might abandon that place or form of interaction almost immediately. Do you happen to see or hear someone say something that inspires or excites or angers you? You may decide to strike up an interaction, or you may opt against it for any of a thousand reasons. Maybe you talk for hours or days. Maybe you cut off all contact with someone because you had a negative interaction with that person. Exciting and all-consuming early days may give way to something mellower. All of this and more happens in face-to-face relationships, in love, and on social media.

Social Media for Connection and Identity Development

While some might view social media as superficial or distracting, professional practices possible through Twitter involve developing an identity, managing impressions, sharing resources, and communicating with other scholars in one's area of study (Veletsianos, 2012). Modern-day internet-mediated communication in social networks involves "Being seen by those we wish to be seen by, in ways we wish to be seen, and thereby engaging in identity expression, communication and impression management are central motivations" (Tufekci, 2008, p. 21). Writers like Maria Popova of Brain Pickings (brainpickings.org) serve up well-written, beautifully-laid-out, easily-shared bites of literary analysis for free, for audiences of hundreds of thousands. Professional organizations and scholarly writers regularly share articles that might otherwise be behind a pay-wall. Thanks to Twitter and Facebook, I have engaged in thought-provoking discussions and debates with people that I have never met, including students at the university where I teach. Social networking is widely accessible and serves as a relatively low-cost way of developing one's connections and seeking out professional development.

Technology and Social Media for Teachers

Options for use of technology for teaching are seemingly endless and ever-evolving. While these range from computer literacy and methods of sharing images and video in the classroom to interacting with students and parents within a Learning Management System (LMS) such as Canvas or Blackboard, this text will focus on the ways that preservice, novice, and practicing teachers can and do use social media sites, especially as they work to develop professional identities as teachers. Furthermore, the ways that scholars in the field of education (including secondary content areas such as social studies and English) use social media affects others' opportunities and thinking. There are findings that suggest recent best practices for these groups as well.

The body of research on social media in education is young and evolving. There have been large-scale studies of how educators use Twitter (e.g. Carpenter & Krutka, 2014), but fewer studies of how individuals think about and understand their interactions on social media, as is called for in Greenhow and Gleason (2012). The research at the heart of this text aims to add to the knowledge base for the latter.

USING THIS TEXT

This text is intended for teacher educators, preservice and inservice teachers, and administrators who use (or would like to begin using) social media in their personal or professional life and work. Even if a teacher would prefer to completely avoid the world of social media for any reason, using social media has become a way of life for millions of people across the globe and will likely remain so for the students who will populate that teacher's future classrooms. Knowledge about how and why teachers and students use social media will enhance any new teacher's options in terms of planning, instruction, and interactions with students and colleagues.

This text has a secondary function: it is also designed to demonstrate the process of conducting action research to determine the effects and impact of one's teaching practice. The text features a blend of theoretical and practical support, including scenarios and case studies based on research with preservice teachers. While books that blend theoretical and practical knowledge for teachers may not usually share their study methods, this text does so in order to demonstrate one way to collect and analyze data for an action research study.

Of course, the reader may start at the beginning and move through the book until the end, but there are various features included to encourage excursions into and higher-level thinking about the content. Several chapters conclude with a summary

CHAPTER 1

to allow for skimming to revisit the material at a later time. There are extension activities at the end of most chapters to extend the reader's consideration about the content of that chapter. These extension activities also guide the reader through the process of conducting a mini action research project, from designing a plan for the research, to collecting and analyzing the data, to writing up the findings for an audience.

The general organization of the text is as follows: (a) a review of literature and practical details about social media and its various uses; (b) methods of the primary study on which this book is based; (c) findings of the study, including case studies of the four preservice teacher participants; (d) a discussion of the aforementioned cases and other research findings; and (e) a conclusion with recommendations for further research and implications for teachers and teacher educators.

CHAPTER 2

RESEARCH SUPPORT AND CONCEPTUAL FRAMEWORK

RESEARCH SUPPORT

The claims and recommendations found in this text include selected literature and current findings on identity, social media, and teaching practices of practicing and preservice teachers; research conducted by the author over more than six years with more than 300 preservice teachers; and, more specifically, the findings of a study of preservice teachers' responses to the use of Twitter and other social media. The bulk of the findings of this study appear in Chapters Ten through Thirteen; however, relevant findings that support the content of Chapters Two through Seven appear in those chapters alongside findings from the literature.

The study had three guiding questions: first, how do preservice teachers respond when required to tweet professionally for a teacher education course? Second, how do these participants use tweets to construct identities, if at all? And third, how do young adult preservice teachers respond to the idea of using Twitter to develop identities based on seeking and sharing professional knowledge?

The mixed-methods data for this study consisted of an electronic survey of preservice teacher candidates who have tweeted for professional purposes in a teacher education course, and case studies that included the public tweets and in-depth interviews with a purposive sample of four survey respondents who represented a range of responses (positive, neutral, and negative) to the class tweeting requirement. A more extensive explanation of the methods appears in Chapter Nine.

CONCEPTUAL FRAMEWORK

Various concepts from the literature converge to provide the framework for the primary study in this text. A key aspect of the conceptual framework that undergirds this study is the idea that using Twitter professionally levels the playing field in terms of whose voice is heard and who does the work of constructing teachers' personae and identities.

Persona and Identity

Other aspects of the conceptual framework are drawn from research on persona and identity formation (Figure 1). Persona refers to the ways that an individual presents herself in public; "*persona* means *mask;* it is defined as 'an individual's social façade or front that…reflects the role in life the individual is playing' ("Persona," as cited in Davis, 2010, p. 2). Furthermore, as noted in Davis (2010):

> Goffman (1959) developed his theory of the Presentation of Self in Everyday Life, which likens people in society to actors onstage. He states that, "when an individual appears before others, he knowingly and unwittingly projects a definition of the situation, of which a conception of himself is an important part" (p. 242). People—teachers included—adopt daily personae based on their and the audience's expectations of the setting. This dramaturgical view of social communication, which includes such features as speech, language, clothing, and gestures (Brissett & Edgley, 1990), provides the basis for the idea that teachers present a persona or play a role onstage in their classrooms; part of that role comes from personal models of teaching, whether fictional or real. (p. 3)

Identity results after personae are adopted in various settings over time (Davis, 2010). Zembylas (2003) has noted that identity for teachers, or one's "teacher self," is not static; it is socially constructed from the power relations and emotions that teachers enact in their school contexts. Gee (2000) adds the element of the audience, who recognizes an individual as a certain "kind of person" in various contexts (p. 99).

Persona and Identity for Preservice Teachers

Issues of persona and identity are especially salient for preservice and beginning teachers enrolled in traditional teacher education programs. During their own schooling, teaching training, semester-long practicum experiences in local schools, and full time internships (often referred to as student teaching), messages about how teachers can and should behave in their roles can be confusing. Representations of teachers in the media can further complicate these ideas about teaching. Teachers may appear in movies such as *Dead Poets Society*, *School of Rock*, or *Bad Teacher* and be depicted as wildly charismatic and passionate, inappropriate, or lazy and unmotivated. Although it can be confusing and challenging, preservice teachers must consider the ways they present themselves and their personae publicly (Kagan, 1992; Lortie, 1975; Weber & Mitchell, 1995); this constellation of factors contributes to a preservice teacher's identity (Davis, 2010). A graphic representation of these factors appears in Figure 3.

The process outlined in Figure 3 illustrates the ways that four preservice student teachers developed and presented personae in their student teaching placements (Davis, 2010). Aspects of the *personal context* included upbringing, prior experiences of schooling, and teacher training. *Classroom and school contexts* such as course goals, instructional levels and *social interactions* with K-12 pupils,

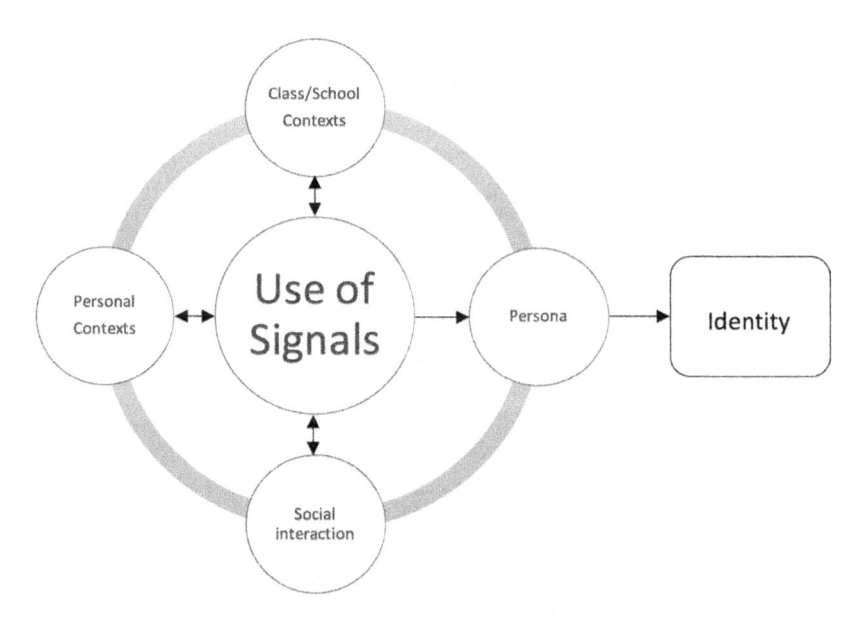

Figure 3. Preservice teachers' development of persona and identity (adapted from Davis, 2010)

mentor teachers, and family and friends were also important to the participants' thinking; these three broad areas factored into the ways that preservice teachers used non-verbal and verbal signals to establish *personae* in the classroom. In turn and over time, these personae became a part of these preservice teachers' emerging *identities*. That is, by speaking to and interacting with students, colleagues, mentors, family, and friends in various situations, and by reflecting on how and why they made their choices about interactions, preservice teachers established their identities as beginning teachers.

While practicing teachers will place differing value on certain influences, and their contexts will change over time—for example, the personal context may expand to include their own children and families in addition to those from their upbringing—all of these areas are still relevant to their identity development. Especially as experienced teachers may refine their knowledge of a school's or district's context, a grade-level curriculum, or the students and families they teach, they also refine the roles they enact in subtle ways.

The research in this section considers how teachers develop personae and identities in face-to-face interactions, but social media provides another opportunity to develop and share these roles publicly. Goffman (1959) has also been a framework for other studies of online interactions, including Davies' (2012) "Facework on Facebook as a New Literacy Practice."

CHAPTER 2

Social Networking Uses and Interactions

Building on the aforementioned concepts, a second key aspect of the conceptual framework is research that has investigated both the purposes for which users employ social networking and the ways in which people interact in these settings, particularly in terms of time (Figure 4). As of June 2016, Prensky's (2001) article coining the terms "digital natives" and "digital immigrants" had been cited in the literature more than 14,000 times. Prensky (2001) argues that today's students have grown up with technology, and so they are learners who think differently from previous generations. These terms have been questioned by some scholars; importantly, as summarized by Ng (2012), these claims have not yet been supported by empirical research.

Expressive, Informational, Visitor, and Resident

Purposes for social networking include the expressive Internet (uses related to presenting a self) as opposed to the informational Internet (uses related to gaining knowledge) (Tufekci, 2008). In other words, one might use Twitter to share articles that will lead others to view that individual as a scholar or leader, or one might use Twitter to locate and read articles about certain topics in order to learn more about them. The questions at the heart of this text reside at the intersection of the two; that is, how do preservice teachers respond when directed to form identities (expressive use) based on seeking and sharing professional knowledge (informational use)? Of course, both purposes may occur at separate times or even simultaneously. Additionally, Wright, White, Hirst, and Cann (2014) describe a range of time investment from digital visitor to digital resident; this study also looks at the ways that participants navigate these roles. Figure 4 offers a graphic representation of the theoretical framework as drawn from these two sources.

As illustrated and described above, the use of social media has various purposes and various levels of investment. Asking a question of a wide audience and obtaining several answers from people all over the globe, as is possible on social media sites, can be wildly exciting from a communication standpoint. However, these opportunities also present challenges. While this process can be viewed as empowering for both the questioner and the responder, it can also be problematic as viewed from a quality control perspective. Is the information or response accurate? Does it reflect research-based best practices that will serve pupils well, or is it based only on the anecdotal experiences or preferences of the speaker?

Third Space

The term "third space" (Bhabha, 1994) has been used to describe a place that combines or lives between the worlds of home and work; this concept has been applied to learning to teach, where the two primary spaces are the college and the practicum setting (e.g. Cuenca et al., 2011; Williams, 2014). Just as the world of the

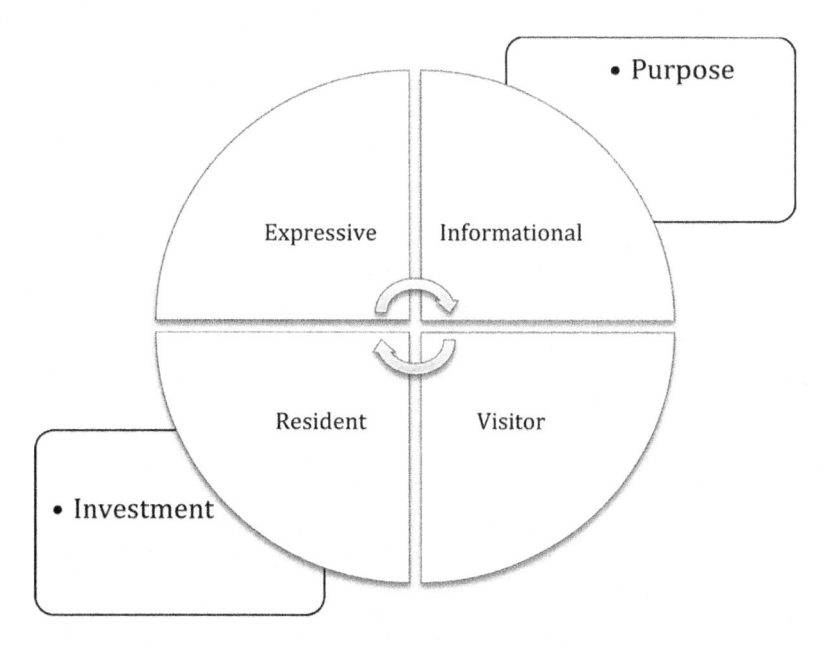

Figure 4. Uses and purposes of social media as based on Tufekci (2008) and Wright et al. (2014)

university and that of the public school can overlap in physical and mental spaces, so too can social media become a third space for teachers to develop identities and share knowledge; it is a place where "interactions can be extended or take on different qualities from the 'official' spaces of interaction" (Davis, 2014, para. 3). The potential of social media is dramatic, but users employ it for varied purposes. Additionally, they may have legitimate concerns about privacy. Use of online communication for purposes such as reflection can be personally transformative, but transformation cannot be imposed (Veletsianos, 2011). When teachers ask students to reflect online, does it lead to change or does it just present the illusion of change? Does online sharing do more harm than good? This text begins to chip away at these and other related questions.

SUMMARY

- The study in this text used a survey and case studies to determine how preservice teachers responded when required to the use of Twitter to construct identities in a teacher education course.
- The study used a conceptual framework that considered how preservice teachers draw on personal and school factors to develop and present personae and identities.

CHAPTER 2

- A second aspect of the conceptual framework included uses of social media; these involved expressive or informational purposes, and a range of time investment from visitor to resident.

EXTENSION ACTIVITIES

1. Make a list of the questions you have about learning as it occurs in your classroom. The following are some possible areas to guide your thinking:
    a. How do students use technology and/or social media for learning and communication?
    b. Which teaching methods do students seem to prefer or find most effective, and why?
    c. What teaching strategies could be added, increased, or decreased, and why?
2. Review the literature to determine what concepts are at the heart of your own research questions and interests. Define key terms and identify their origins in others' research and writing.
3. Construct a graphic organizer to illustrate how these concepts work together.

CHAPTER 3

SOCIAL MEDIA

A Comparison

By providing multiple opportunities for selective self-presentation—through photos, personal details, and witty comments—social-networking sites exemplify how modern technology sometimes forces us to reconsider previously understood psychological processes.

(Gonzales & Hancock, 2011, p. 82)

Simply the act of committing the names of social networks to print seems slightly foolish, as the details and even existence of the following will certainly change with time. Nevertheless, there are some basic considerations of different social media platforms that, when compared, offer a broader view of the purpose of these forms of computer-mediated communication (CMC). While various organizations and individuals have compared the uses of social media, especially for marketing purposes, Table 1 considers various aspects that might directly affect preservice and practicing teachers as they choose among these sites.

Table 1 outlines stability of connections, anonymity, resource sharing and depth of interactions. The term *stable connections* is defined as whether the user can maintain a list of people or accounts that he or she follows, and then revisit these accounts from time to time. Friends or followers may change, but, in general, these connections can be retrieved. *Anonymity* is whether a user can post or comment from an account that is not his or her real name. *Resource sharing* refers to the kinds of links that can be shared (generally text, links to websites, or images in the form of photos or videos). Finally, *depth of interactions* is certainly widely variable on any given site, but is judged based on the amount and quality of communication with identifiable individuals over time through the site. For example, Facebook is known as a place where people share images and text (without a 140-character limit, as exists on Twitter) and comment on their friends' posts; hence, the depth of interactions possible is high, even if not all members of Facebook use it in this way.

One may act as a resident or visitor (Wright et al., 2014) one each of the above platforms, but in sites with a high depth of interaction such as Facebook or Twitter, it may be more likely for one to become a resident. Likewise, sites may be used expressively or informationally (Tufekci, 2008), but ones that discourage anonymity and encourage resource sharing, such as Twitter, may yield more individuals who seek both uses. A site such as Yik Yak has different purposes, and users cannot express a consistent identity because it is anonymous and connections with others are

CHAPTER 3

Table 1. A comparison of social media sites

	Stable Connections	Anonymity	Resource Sharing	Depth of Interaction
Facebook	Yes	Discouraged	Yes	High
Instagram	Yes	Possible	Yes	Medium
Pinterest	Yes	Possible	Yes	Medium
Yik Yak	No	Yes	No	Low
Slack	Yes	Possible	Yes	Low to High
Twitter	Yes	Discouraged	Yes	Low to High
Tumblr	No	Yes	Yes	Low

not maintained—one's avatar is not a photo and changes with each new interaction. Sites such as Tumblr do allow users to express identities, but some choose to do so anonymously (Reeve, 2016).

The uses of social media in education have been studied, to include Twitter (e.g. Carpenter & Krutka, 2014), Facebook (e.g. Manca & Ranieri, 2013), and Instagram (Milton, 2015). Teachers may use sites such as Pinterest during the planning process for resource collection and sharing as a kind of virtual bulletin board; one glance at a search of "teaching ideas" yields hundreds of various image-based "pins" on the site.

DETAILS ABOUT SOCIAL MEDIA USE

Connections

Numbers from surveys vary, but one source suggests that 65% of adults in America use social media (Pew Research Center, 2013). These numbers have been steadily growing in all age groups for the last decade; surveys show that the age group with the highest percentage of social media users is made up of adults aged 18–29 (Pew Research Center, 2013).

As seen in Figure 3, a majority of adults in nearly every age group use social media. For ease of interpretation, the lines on the legend follow the order of the final results: 90% of users aged 18–29 used social networking, and 46% of users over 65 years old used social media.

Such widespread use of social media suggests that former methods of reconnection may now be obsolete. The surprise is gone. Recently I experienced a wedding where I was lucky enough to interact with people that had not seen or spoken to in more than a decade—many of them were not active on social media, or were friends of friends and so we had lost touch. We caught up on all the major events that had occurred; there had been a kind of mystery about what each person had been doing and whom they had become. In the past this kind of reconnection might have occurred more often at events such as high school reunions, but for my

Social networking site use by age group, 2005-2013
% of internet users in each age group who use social networking sites, over time

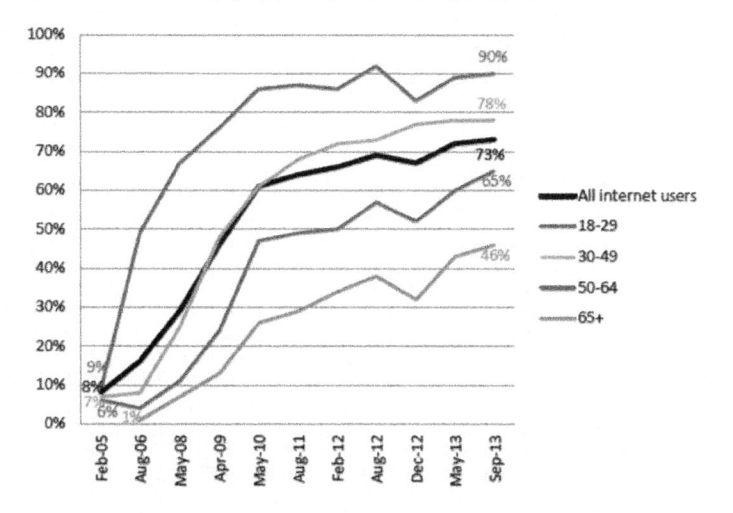

*Figure 5. Social networking site use by age group, 2005–2013
(Pew Research Center, 2013, para. 5).*
Source: Latest data from Pew Research Center's Internet Project Library Survey,
July 18–September 30, 2013. N = 5,112 internet users ages 18+. Interviews were
conducted in English and Spanish and on landline and cell Phones. The margin of
error for results based on internet users is +/– 1.6 percentage points.

generation and later generations—born in the 70's and later, and spreading to older adults as well—we use social media to connect with others around shared interests. Long gone are the times when people fell out of touch with high school classmates, only to catch up with them in person twenty years later. New college students about to meet their roommate for the first time do not need to resort to letters or the phone: they can locate their prospective roommates on social media and view images and maybe even videos of them immediately (personal communication, V. Fantozzi, June 15, 2016). Now, for most people, almost any former or future classmate can be located on social media and remain frozen in profile form there in cyberspace for impromptu investigations whenever the inspiration strikes.

Communication

Sites such as those outlined in Table One offer various ways of interacting with other users. Typically, this takes the form of comments or other means of recognizing posts ("likes" or "hearts"). Teens communicate with friends and share information on sites such as Facebook, Instagram, and Twitter. Social media is not solely about posts and comments, however; many sites also includes messaging features that allow

CHAPTER 3

those who are connected to each other to communicate directly through a private message. Postings on social networks have even developed their own syntax in some cases, and in places like Tumblr and Facebook, messages are sometimes deliberately "rendered in the uncapitalized and unpunctuated casualness of instant messages" (Reeve, 2016). Punctuating a post or message, especially a text message, with a period can be viewed as harsh or insincere (Gunraj, Drumm-Hewitt, Dashow, Upadhyay, & Klin, 2016). Other common strategies involve responding with or initiating posts with gifs, which are short looping videos that can be trimmed from longer videos such as movies; memes, which are images with overlaid text that generally comments on the feeling or message of the image; and emojis, which are small clip-art style pictures that can be added to a longer post or used alone as a response. Emoji might include a picture of a vomiting face or a yellow cat face with hearts for eyes; multiple emoji may be used at one time and appear as a string of images.

As of June 2016, the popular account Common White Girl (@girlposts) had 62,300 tweets and more than 6 million followers. The account uses images, gifs, and emojis to illustrate and punctuate comments and observations about crushes, popular culture, and interpersonal relationships. The account does not represent one individual and her life, but rather appears as a representation of interests that, as noted in the title of the account, might interest and even represent others in a certain demographic.

Figure 6. Tweet with emojis @girlposts in response to a retweeted, complimentary tweet (Common, 2016)

When an organization of which I am a member began using Slack for group conversations, I found myself suddenly obsessed with finding the most appropriate emoji to punctuate my friends' comments. Research at this point is limited, but some studies have investigated how teens use emojis and found that they are more likely to use them in social-emotional contexts than in task-oriented contexts (Derks, Bos, & Grumbkow, 2007). College course instructors have begun using Slack for course-related conversation and noted that it can help build community and add a sense of fun and humor through such integrations as "/cat facts" (Whalen, 2016, para. 28).

Identity

It is far less common to think of social networks as a place where persona and identity are developed, although ironically it was more evident in the name of an older site: MySpace. Originating in 2003 and evolving over more than a decade, MySpace was an online site where people could build a profile and interact with friends and strangers.

In her book *It's Complicated: The Social Lives of Networked Teens*, danah boyd (2014) explored the ways that teenagers live with technology. While, as boyd has noted, parents may fear the interactions that teens have online, teens will find ways to use social media wisely and as a release and tool for identity development. Additionally, adults and kids alike gravitate to certain social networks for different reasons. Reeve (2016) described the possibilities of various networks, especially among teenagers:

> Each social media network creates a particular kind of teenage star: Those blessed with early-onset hotness are drawn to YouTube, the fashionable and seemingly wealthy post to Instagram, the most charismatic actors, dancers, and comedians thrive on Vine. On Facebook, every link you share and photo you post is a statement of your identity. Tumblr is the social network that, based on my reporting, is seen by teens as the most uncool. A telling post from 2014: "I picked joining Tumblr and staying active on here because: (1) I'm not attractive enough to be a Youtuber (2) Not popular enough for Twitter (3) Facebook is dumb." You don't tell people your Tumblr URL, you aren't logging the banalities of your day—you aren't even *you*. On Tumblr, you can revel in anonymity, say whatever you want without fear of it going on your permanent record. (para. 6)

Reeve (2016) later shares an observation by Jason Wong that, "teens perform joy on Instagram but confess sadness on Tumblr." If young people and adults alike use these many sites to share various aspects of their lives and, ultimately, their identities, how do they determine what to share and how to share it? The following section discusses some of the considerations that emerge when sharing and interacting in these settings.

BLURRING THE LINES

When we share personae, whether in person or online, that do not feel authentic or multifaceted, it can lead to negative feelings or a desire to present a more complete self. Ideas of persona and the complications that arise from it extend to philosophy, social psychology, and the arts, including writing. Khalil Ghibran (1883–1931) described the experience of feeling as if he were without a mask:

> For the first time the sun kissed my own naked face and my soul was inflamed with love for the sun, and I wanted my masks no more. And as if in a trance I cried, 'Blessed, blessed are the thieves who stole my masks.'
>
> Thus I became a madman.

And I have found both freedom of loneliness and the safety from being understood, for those who understand us enslave something in us. (as cited in Popova, 2016)

When Ghibran writes, "those who understand us enslave something in us," it calls the mind the feeling of being predictable or of having others expect us to act and look in certain ways. Many of us can and do give these audiences what they expect, but it can be stifling.

What are the lines, and how and why would one blur them? To put it simply, the lines when it comes to social media are those that exist between one's personal life and one's professional life and their associated audiences. These are further explored in later chapters, but at this point in the text it will help to envision what separations one might create, and why.

An Example of Line Blurring

My own Pinterest page features boards for each of my children; clothing styles that I like (to share with a popular personal styling service); parties that I have planned; lessons that I find interesting; images that I share as part of my teaching; and resources that I share as part of my research, such as links to interviews or articles about oral history or digital storytelling. Pinterest is an example of a place where I blur the lines between my personal and professional lives because I judge the stakes to be low. The content is only images, I do not comment on the images beyond perhaps a few words, and the purpose of the site is different. I do not use Pinterest to develop my identity or public stance on a topic, as I do on Facebook (where I connect with my friends, family, and former students, and share details about my personal interests) or Twitter (where I connect with scholars and students and share resources and details about my professional interests).

For me, there is very little overlap between Facebook and Twitter; when it occurs, it may be because an interest group I follow is more likely to post certain kinds of information on one site (Twitter is generally better suited for sharing real-time updates such as game scores). I also have an unofficial policy that once my former students graduate, if they ask to befriend me on Facebook, I accept; I count at least 20 former secondary and college-level students among my Facebook friends and interact with them to varying degrees (some not at all, some as often as a comment or "like" once a week). My college-level students, on the other hand, often must follow me on Twitter (I follow them too) and we use a course hashtag for certain courses; these interactions are a small part of their grade. It was the student response to this course requirement that inspired some aspects of the primary study that appears later in this text.

As we share resources, communicate, and explore online groups and people that interest us, we are enacting personae. We are being seen by certain people and in certain ways that we wish to be seen (Tufekci, 2008). Together these personae create a public identity that may be personal, professional, or some blend of the two. We may have separate public identities in separate spaces.

Blogging is one way that many teachers have begun blurring the lines between their personal and professional identities on social media. New and veteran teachers alike may blog about challenges or successes in the classroom or beyond. At least one teacher blog features a section devoted to ideas that she has tried and that have failed. Compare this strategy to a scholar online who only posts or blogs about major public successes: the former has constructed a multi-faceted, relatable public identity; the latter has created a one-dimensional representation of herself that, one suspects, must feel less satisfying or authentic.

For Modeling

Maria Popova (2016) describes her work on the Brain Pickings website as "a record of my own becoming as a person—intellectually, creatively, spiritually—and an inquiry into how to live and what it means to lead a good life" (para. 2). A use of social media that has emerged in my own teaching is similar to Popova's; I use Twitter to model professional interactions with students, scholars, and colleagues. Images and articles abound in my Twitter feed. A teacher or college professor can describe a conference to a group of students, but the idea may seem difficult to grasp without the images of a convention hall filled with books and other teacher resources, videos of teachers scurrying to create sculptures out of hastily scavenged promotional materials in a maker lab session, or photos of slides from scholarly presentations in one's field. When I tweet a light-hearted observation about a painting to the Art Institute of Chicago and the museum tweets back, or when I retweet author Mary Karr's beautiful writing environment with the hashtag #writinggoals and she "hearts" it, my students who follow me on Twitter can observe the ways that people and organizations in the world come together around issues large or small, simple or controversial, and fascinating or mundane.

Figure 7. Retweet of author Mary Karr's tweet with added hashtags (Davis, 2016)

CHAPTER 3

Within or Across Formats

Using social media for personal interactions tends to occur before professional interactions. One survey shows that 70% of university faculty use social media in their personal lives, while about 55% use it professionally and slightly more than 40% use it in their teaching (Seaman & Tinti-Kane, 2013, p. 8).

In the process of developing various online selves, one must eventually ask: who am I in each of these settings, and why? Are there lines between my personae, and if so, why? If I am blurring the lines between personal and professional, which lines am I blurring? Why? Across sites or within sites? Can or should these selves really be separated? What functions are my masks serving—are they protective, or are they providing confidence that is necessary or unnecessary (as with negative comments on discussion boards)?

These considerations raise the question: could some frustration with social media be due to a misunderstanding of their purpose and how or why these sites can be used? In addition to these potential considerations about identity and persona, there are several more straightforward challenges that affect users, which are described in the following section.

CHALLENGES

Anyone who uses social media will encounter various challenges. The following section will focus more specifically on the complications confronting teachers and students alike when the world of social media bumps up against the world of school.

Access/Perception

A common complication with social media for teachers and K-12 students is that districts often block these sites, especially for students. One survey found that 34% of school-based respondents said that "their district[s] blocked the site for students only," while "15% blocked both students and teachers" (Carpenter & Krutka, 2014, p. 425). Some districts have begun to institute Bring Your Own Device (BYOD) policies to increase student use of the internet during school time, but there is wide variation in how these policies are utilized. Additionally, some teachers find the use of cell phones in classes distracting or disruptive. Beyond the simple fact that many cannot access certain sites at their home or place of work, there are other negative perceptions about social media; some of these perceptions include that those who use it are frivolous or conceited.

Frequency of Use/Adoption

Whether or not students use social media and which sites they choose can be linked to their friends' uses of those sites. For teachers, school administrators can influence how and whether social media is used, especially to document professional practices such as in-class projects and student work. The superintendent's view may not always be a consideration unless it is especially positive or negative, but in districts where the superintendent and district are supportive and encouraging of social media, teachers are more likely to use it for professional purposes (Edutopia, 2015). In Albermarle County, Virginia, teachers can earn professional development credits for actions such as producing blogs or other social media content or participating in educational chats (Edutopia, 2015). In these few districts and schools that actively support work on social media, including my own daughters' school, one might see teachers' tweets about lessons, images of students on field trips or at work on projects, or photos of student work samples. However, those concerned about obtaining a first teaching job or maintaining job security may have legitimate concerns about jeopardizing these through the use of social media if it is not prevalent among teachers in the district.

Time is also a commonly cited area of concern for teachers. Forming and interacting with a Professional Learning Network (PLN) takes time, which can be in short supply for teachers.

Comfort with Technology

When access to technology or the time to use it, whether at home, school, or both, is limited, users may experience challenges because they have had less practice and experience. Teaching is a profession with a wide range of technological skill and levels of access to the internet, so some teachers may fear that they are not skilled or knowledgeable enough to use social media properly. As seen Figure 5, social media use dropped as age increased; the only group with fewer than 50% using social media was comprised of adults over the age of 65. The teaching workforce today includes adults of all ages, but because computers were not a regular part of life for older adults as they have been for younger adults (personal computers were introduced in the 1970's and 80's), it stands to reason that social media may not be an interesting, relevant, or even accessible aspect of life for some teachers.

Respecting Boundaries

Introducing social media in the classroom and into teachers' and students' lives means that some parties may feel uncomfortable about connecting in these ways. The phenomenon known as the "creepy treehouse" (Stein, as cited in Young, 2008) occurs when students view a space as uninviting or scary simply because a teacher has moved into it and is encouraging young people to connect there; students generally think of social networks as places where they can socialize with peers, not with teachers. These thoughts may stem from simple preferences for interaction with peers, or perhaps from a darker place of prior warnings about connecting with strangers or adults in online spaces.

CHAPTER 3

Privacy

Finally, one of the most common challenges for teachers at college and K-12 levels is privacy. After "integrity of student submissions," the second most-cited barrier to faculty use of social media in a recent survey was "concerns about privacy" for both faculty and students (Seaman & Tinti-Kane, 2013, p. 17). Myriad cautionary tales—some of them described in this text's foreword—of teachers who were tagged in pictures with alcohol or in skimpy clothing which then led to a firing have now made their way into teacher education discussions and recommendations. Universities have sanctioned or even suspended or expelled students for troubling posts on social media.

SUMMARY

- Different social media platforms offer different opportunities for users.
- A majority of adults in America use social media; for young adults the current figure is about 90%.
- Identity can be developed on social media; whether or how to blur lines between personal and professional selves is often a consideration for teens and adults.
- There are several challenges associated with using social media, to include access, comfort, boundaries, and privacy.

EXTENSION ACTIVITIES

1. Consider and reflect on how and why you use social media in your everyday life.
2. Imagine that a former high school teacher encouraged you to connect on social media when you were in high school. How would you have responded?
3. Develop a personal strategy for how and why you will connect (or avoid connecting) with K-12 students when you are a teacher.
4. Return to your research topics of interest from the Chapter Two extension activities. Review the literature to define key terms or to identify related concepts.

CHAPTER 4

PROFESSIONAL DEVELOPMENT OPPORTUNITIES ON SOCIAL MEDIA

> Learners become amplifiers as they engage in reflective and knowledge-building activities, connect and reconnect what they learn, add value to existing knowledge and ideas, and then re-issue them back into the network to be captured by others through their PLNs.
>
> (Warlick, 2009, p. 16)

Professional development occurs on social media sites such as Twitter in a wide variety of ways. Educators at all levels connect online with others in their field (Carpenter & Krutka, 2014; McCarroll & Curran, 2013; Warlick, 2009). In these online settings, educators share resources and participate in education chats with goals such as "combating isolation and finding community" (Carpenter & Krutka, 2014, p. 428). Popular hashtags such as #edchat organize conversations across participants who may not be connected to one another. The hashtag provides the connection.

Use of Twitter and similar social media sites involves considering more than just sharing links or thoughts; savvy users consider a wide variety of methods and forms of sharing information. Aarts, van Maanen, Ouboter and Schraagen (2012) found that tweets have characteristics of the actor, message, and network. Past research has viewed actor characteristics to include individuals' informational "influence and conformity to social pressure;" message characteristics are related to the arousal that the message has on audiences; and network characteristics consider what about various networks lead to the spread of ideas (Aars et al., 2012, p. 739). The aforementioned categories may actually be artificial and should be viewed together instead of separately (Aarts et al., 2012). Tweets with certain features such as hashtags, URLs, and use of a username @ are more likely to be retweeted (Naveed, Gottron, Kunegis, & Alhadi, 2011; Suh, Hong, Pirolli, & Chi, 2010), thus expanding the original author's message and influence. PreK-12 and higher education scholars now form Professional Learning Networks (PLNs) in order to communicate and share resources with others in their field.

Investment of time and ease of use figure into the commonly cited "digital native" or "digital immigrant" (Prensky, 2001) and the more recent and statistically supported "digital resident" or "digital visitor" labels in Wright et al. (2014). Additional research needs are varied, including why chains of social influence remain short in online situations (Aarts et al., 2012). Twitter can serve as a network of scholarly learners (Veletsianos, 2012), and preservice and novice teachers are engaged in the process of becoming scholarly learners. Veletsianos (2012) found the following common themes in his qualitative study of how higher education scholars used Twitter: "information, resource, and media sharing; expanding learning opportunities beyond the confines of the classroom; requesting assistance and offering suggestions; living social public lives; digital identity and impression management; connecting and networking; and presence across multiple online social networks" (p. 7).

BUILDING A PROFESSIONAL LEARNING NETWORK

Building a PLN is at the core of successful online interactions. When one's connections are extensive, it will allow for richer and more diverse posts and resources in one's feed. Beyond simply accessing resources, these networks can impact one's opportunities, knowledge, and identities, even when one may be geographically isolated. As Gee (2000) noted:

> Wealth and power tend to stem from whether or not one has access to specific networks of people and information spread across the country and the world and to specific experiences connected to these networks. In turn, these networks and their concomitant practices allow people to form multiple, changing, and fluid A-Identities with others, some of whom they may see in person but rarely. (p. 121)

Gee's (2000) term "A-Identities" refers to the affinity groups to which one belongs and the ways that those groups contribute to the identities we form; a more detailed explanation of Gee's forms of identity appears in Chapter Eight. Connecting with strangers across the country and world was once exceedingly rare and complicated, but social media offers this chance to any teacher who has internet access and the desire to connect.

The idea of connection can be simple but for some teachers social media such as Twitter can be vast, stressful, and intimidating. Teachers, scholars, companies, and strangers everywhere are posting everything from links to lesson plans to images of students at work in the classroom to videos of their pets reacting to adorable costumes. In order to make professional development on social media manageable, one must let go of the desire to keep up with all tweets or even people or organizations, and simply check the feed of posts as often as is possible. Martha Burtis, Director of Digital Knowledge Center at the University of Mary Washington and the author of this text's foreword, refers to this as "dipping into" the never-ending stream, which can allow for playfulness and serendipitous discoveries and interactions (personal communication, February 7, 2013). Some teachers may carve out a larger amount of time for social media, but ignoring it entirely would be ignoring research findings and recommendations from various professional organizations that it is a useful method of connecting with other professionals. A simple way to ensure that this feed is well-stocked with stories and comments that relate to your interests is to follow organizations, teachers, and scholars in the areas that interest you. There are other methods to increase your use of Twitter or other social media, to include social media management tools such as Hootsuite or Tweetdeck (pictured in Figure 8) that display continuously updated searches for hashtags or specific users.

Figure 8. Example of tweet compiling tool as shared in a tweet to students (Davis, 2013)

The above photo also features the use of course hashtags, which are described in the following section.

Group Chats and Hashtags

Searching by hashtag is a simple way to view the posts of others about similar topics, even if those people do not follow or even know each other. Similarly, group chats occur according to a predetermined schedule, are organized around shared professional interests, and utilize hashtags to facilitate conversation among the participants of the chat. Seventy-three percent of the Twitter-using educators who were surveyed by Carpenter and Krutka (2014) indicated that they used Twitter to participate in chats for professional development (p. 473). A Google search of education hashtags and chats will yield several suggestions; at the moment, one of the most commonly cited is #edchat, but given the ephemeral nature of social media, it would be wise for educators to search online for the most recent user recommendations.

Resource Sharing

Twitter and Facebook can be especially useful for sharing articles with colleagues and students, as long as the sharer can count on the intended audience seeing the article. Adding a hashtag as described above is a useful way to funnel articles to one location, where they can be seen by anyone who chooses to follow that hashtag.

CHAPTER 4

Some organizations such as the National Council of Teachers of English (NCTE) utilize conference hashtags such as #NCTE15 or #NCTE16. They even enlist various members of the organization to tweet during different times of the day and during different sessions (M. Davis, personal communication, November 14, 2013) and assign each session its own hashtag for attendees who wish to have back channel conversations. For other conferences and organizations, live tweeting in this manner may happen organically among people who happen to use social media either personally, professionally, or both. As someone who has live tweeted during a conference, I will add that one effect of the practice can be be involving one's audiences at the expense of making one's own physical attendance and participation in the live session more distracted and fragmented.

Retweeting and Remixing

Retweeting is simply re-sharing someone else's post or images on Twitter; this is also possible and even encouraged through the use of built-in or easily installed buttons on sites such as Facebook (called "sharing') and Instagram (called "regramming"). An example can be seen in Figure 7 in the previous chapter. It is perhaps fitting that the originator of the term "remix culture" is unclear, but the premise is that nothing is original; we are all building on what has come before, and borrowing liberally from others (Ferguson, 2012). Memes are widely shared images or other posts that are often shared through social media. Nearly every university student and Reddit or Twitter user will recognize the standard meme format of an image with white text (see example in Figure 9), but due to the nature of construction and distribution, it would be impossible to identify the original author.

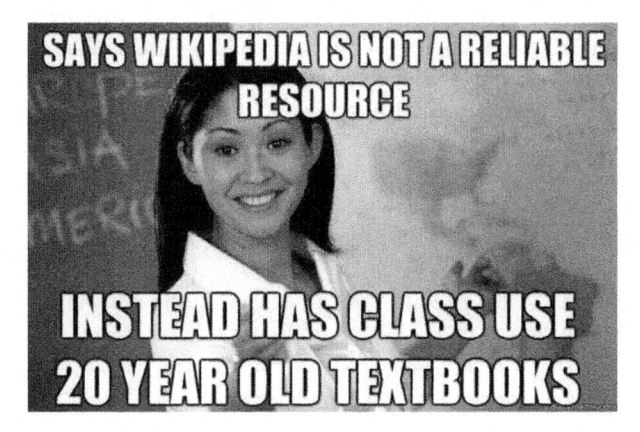

Figure 9. Example of meme, "unhelpful high school teacher" (Unhelpful, n.d.)

The "unhelpful high school teacher" meme is meant to recall frustrating moments from high school; in another version the words note, "can't spell a word? Here, look it up in this book where words are organized by spelling." Sharing such sentiments can be a common form of interaction on social media.

Following and Followers

The following graphic is a framework for various groups and individuals to guide following accounts on Twitter or a similar social media site when used for professional purposes.

Who
- Teachers: local/national/international
- Researchers

What
- Lessons
- Articles/Current Events

Why
- Blogs
- Professional Organizations

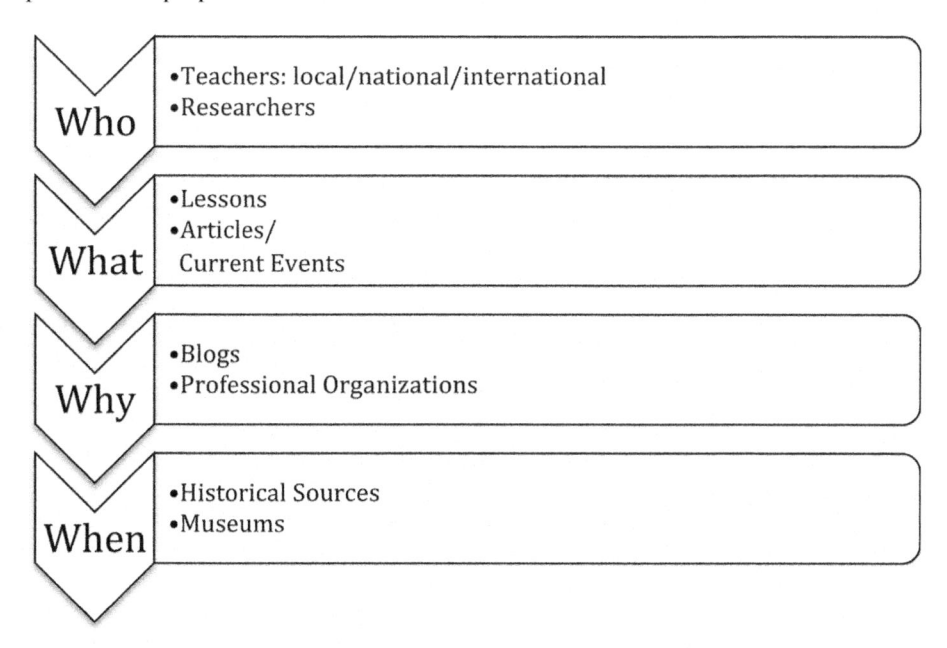

When
- Historical Sources
- Museums

Figure 10. Sources to follow

Figure 10 illustrates the various ways that both preservice and practicing teachers have built their PLNs. Guiding questions include *whom* to follow, *what* to seek, *why* users are sharing this information, and *when* in the past that common curricular elements may have been salient. The following is an example of the search terms that might help a secondary English teacher to build a PLN on Twitter:

- English teacher
- English teacher + location ("English teacher Charlottesville")
- Literacy research
- Literacy specialist
- Shakespeare (turns out he's on Twitter!)
- Favorite author name

25

CHAPTER 4

- Secondary English publication
- Secondary English lessons
- English teacher blog
- National Council of Teachers of English (inquire among colleagues and mentors about recognized professional groups for your content area)
- Teachers of English + location (i.e. Virginia Association for Teachers of English)
- National Writing Project
- National Writing Project + location
- Museum
- Storytelling
- Oral history
- Museum + location ("Museum Fredericksburg")
- Library
- Library + location

This is one example, but of course, these terms will change depending on one's grade level, area of focus, and content-related interests. In practice, I have found that teachers often develop similar numbers of accounts who follow them (followers) and accounts that they follow (following). Twitter uses an algorithm to share recommendations of additional accounts to follow, and by following those accounts, users can build ever-growing networks. An additional strategy to build a large and diverse network is to borrow from the accounts that others follow; I often recommend that my students review the accounts that I follow and follow the ones that relate their own interests. By continually following new accounts, users will gain more followers as those accounts follow them back. A varied and diverse PLN will lead to a more varied feed.

Evaluating Social Media Accounts

Just as readers and writers must review sources for research in everyday life, so too will they need to evaluate the accounts that emerge in a search for social media contacts and resources. Some will be "bots," which post updates or retweet certain posts automatically and are not run by humans in the same way that a personal account might be, while others may be profit-making ventures by companies or individuals. There will be various irrelevant strangers that follow users for all manner of reasons; users can prune these accounts by blocking them. It can be challenging to clear the initial hurdle of connecting to strangers, because often recommendations are to avoid that; however, Twitter is generally a place where people can and do connect to and learn from strangers who may be experts in their fields—and who are also willing to communicate, or least connect, online.

SUMMARY

- Social media offers professional development opportunities for teachers that include educational chats, hashtags, and shared resources.
- Teachers wishing to use Twitter for professional development should form a PLN by searching for others who share their interests and investigating other accounts through their existing connections.

EXTENSION ACTIVITIES

1. Review and borrow at least 20 new sources to follow from other teachers and scholars in your field.
2. Locate an example and a non-example of a social media account operated by a scholar in the field of education.
3. Search Twitter to identify scholarly articles related to your research interests.
4. Review additional literature related to your research topic to determine how other teachers and scholars have considered or addressed these or similar issues.

CHAPTER 5

SOCIAL MEDIA IN THE CLASSROOM

> Our presentations of self, and our relationships are increasingly being enacted through screen-based text-making, and this activity allows us to read our presentations of self and each other in multiple ways. Text-based social networking has become something that people 'do' and for many has become embedded in their quotidian lives.
>
> (Davies, 2012, p. 28)

There is no shortage of warnings about how teachers should not befriend their students on social media or share potentially damaging photographs. Teachers or schools may have policies about the ways that teachers or students can connect online. A survey of educators on Twitter found that a small percentage (16–17%) of teachers use Twitter for in- or out-of-class activities for students, which may be due to the restrictions that schools place on the site (Carpenter & Krutka, 2014, p. 424). Use among university faculty is more common: in a representative sample of nearly 8,000 faculty members, 41% noted that they use social media for their teaching on at least a monthly basis (Seaman & Tinti-Kane, 2013).

There are various measures that might enable teachers to take advantage of the strengths of social media while also protecting against possible issues that might arise in the future. These measures range from the simple (do not share any potentially offensive material) to the more complex (change privacy settings or share links with smaller subgroups). The most dramatic measure that some might take—not participating in social media at all—is not recommended, because doing so would limit options for resource sharing and connection.

One teaching situation where use of social media is more common is online courses in higher education. Major (2015) identified various considerations that teachers in online spaces might make to develop specific teaching personae, such as choosing a user name and using social media. Professors or even high school teachers who teach online do not have the same constant opportunities that face-to-face teachers have to construct personae in the classroom such as clothing choice, mannerisms, and tone of voice, but new options exist (Major, 2015). Considering one's personae and identities may even be simpler and more straightforward when looking at how one presents oneself on social media.

Preservice teachers in a study of persona became better able to see the differences in their selves when asked about their "Facebook self" (Davis, 2010). Helpful questions in the aforementioned study included *Who are you when you post updates on social media, with its host of followers from all areas of your life? Compare that to who*

CHAPTER 5

you are when you are meeting with friends or having lunch with your parents. In a nod to professionalism, one student commented that she changed the settings on her Facebook profile (to make it only visible to her friends, as opposed to anyone who searched for it) and removed tags on photos of her that could be seen as inappropriate—this was a common recommendation from her earlier teaching methods courses—and that this attention to how she presented herself online was "an example of how [she] sort of [had] to be a different person in the classroom" (Davis, 2010).

CALLS FOR USE OF SOCIAL MEDIA IN EDUCATION

While some educators may deride social media or claim that it does not have a place within our classrooms or professional lives, professional organizations such as the National Council for the Social Studies (NCSS) have called for instruction and analysis of how we use technology, to include social media.

The NCSS (2013) position statement argues that media literacy extends to social media and that online environments can help engage students as citizens in a democratic society; however, the statement also advises caution as we use and teach with this technology. The policy states,

> The use of social media to create multiple online and blended economic, political, and social settings with a global reach requires rethinking how to prepare children and youth to participate in such settings. In turn, such settings are reshaping how children and youth are able to act as citizens and consumers. Helping students make sense of all the information, new environments, and ways of being, requires grounding them in the experiences of those in the past and other civic and cultural settings. In a time when as a field social studies struggles for relevance, social studies educators need to recognize and promote how they are uniquely qualified and situated to enable young people to effectively use mobile technologies as a citizen, learner, and member of a democratic society in a global setting and to explore the civic, economic, and social implications of such technologies across time and place.

The above call has even provided a useful guide for classroom action research (Milton, 2015). Additionally, the National Council of Teachers of English (NCTE) has released a similar position on teaching with technology. The NCTE (2013) policy states,

> Literacy has always been a collection of cultural and communicative practices shared among members of particular groups. As society and technology change, so does literacy. Because technology has increased the intensity and complexity of literate environments, the 21st century demands that a literate person possess a wide range of abilities and competencies, many literacies. These literacies are multiple, dynamic, and malleable. As in the past, they are inextricably linked with particular histories, life possibilities, and social

trajectories of individuals and groups. Active, successful participants in this 21st century global society must be able to

- Develop proficiency and fluency with the tools of technology;
- Build intentional cross-cultural connections and relationships with others so to pose and solve problems collaboratively and strengthen independent thought;
- Design and share information for global communities to meet a variety of purposes;
- Manage, analyze, and synthesize multiple streams of simultaneous information;
- Create, critique, analyze, and evaluate multimedia texts; and
- Attend to the ethical responsibilities required by these complex environments.

The benefits noted by the above calls for the use of social media in education include that it supports students as they become global citizens, builds connections with varied groups, and develops digital literacy skills.

COLLEGE-LEVEL USES

Preservice teachers are also college students, and may have experienced the use of social media in courses as undergraduates. Twitter and other social media are often used in college-level courses to increase the audience for assignments, sometimes exponentially. Conversely, Twitter has been used to facilitate small-group discussion environments in large lecture courses (Rankin, 2009). Recommendations from Rankin's (2009) experiment included the instructor's creating small groups, providing discussion topics, and circulating around the room during the online discussion; a benefit is that more students were engaged in the class discussion.

Dabbagh and Kitsantas (2011) referred to a Personal Learning Environment (PLE) on social media and proposed a framework to support college students' self-regulated learning; the framework included three levels: personal information management, social interaction and collaboration, and information aggregation and management (p. 5). For example, when using social networking tools, college students might be encouraged to first create a profile on a site such as LinkedIn, then connect to relevant organizations, and finally to reflect on the experience and process, "to enhance the desired learning outcome" (Dabbagh & Kitsantas, 2011, p. 5).

The use of social media to explore identities occurs at the college level because college students are already engaged in developing their identities in a new setting and are about to embark on future careers. Jeffrey W. McClurken, professor of history & special assistant to the Provost for teaching, technology, and innovation at the University of Mary Washington noted:

Some of the benefits are that they are crafting a digital identity, and the benefits to me are that, when students do things in public, they are more thoughtful than when they're doing it just for me or their classmates. It allows them to

CHAPTER 5

think through these tools in a new way, and to think through the channels of communication that they have in their lives and that they have access to. They have a way to think about how to use those in critical ways, in reflective ways, ways that push them beyond the 7- to 10-pages. (personal communication, January 14, 2016)

Seaman and Tinti-Kane (2013) found that the most commonly used digital formats among university faculty in terms of student creation of material were blogs and wikis, followed by podcasts and social media such as Twitter and Facebook; between 10–20% of the faculty surveyed used Twitter and Facebook for student assignments. The following tweet is an example of one way that a college instructor might use a tweet to share resources such as a blog post with students in a course (in this case, two courses) as well as with program-area colleagues.

Janine Davis @JanineSDavis · Jun 7
getschooled.blog.myajc.com/2016/05/27/uni ... #TOEDavis from our text's author! #351Davis and also thought of you and your classes, @Techtweed and @Jen_D_Walker

Figure 11. Tweet to course hashtags and colleagues (Davis, 2016)

Of course, use of social media for education is not limited to college instructors; there are applications for K-12 teachers as well.

K12-LEVEL USES

Both youth and adulthood are times of time of identity exploration and development (Erikson, 1968), and relationships are now enacted online as well as in-person (Davies, 2012). Young children, referred to as "early adopters," incorporate technology into their play (Wohlwend, 2009, p. 117). Districts and educators can capitalize on students' interest in technology and social media to use it for communication and class assignments; however, any use of social media requires an awareness of potential dangers and concerns among parents and students. Some of the most publicized concerns in popular media currently have been about cyber-bullying and online predators.

Acceptable Use Policies

Many districts develop technology acceptable use policies (AUPs) that guide K-12 students' and teachers' options for technology and social media. In their paper, Ahn,

Bivona, and DiScala (2012) noted the conflicting policy decisions and frames that can affect student learning and teachers' options; these included online safety, a need for technology to keep pace with global competitors, and increased access to technology. In their analysis of large school district AUPs, Ahn et al. (2012) found that about 18% of school districts named information and media literacy as a goal of using technology. Additionally, "many districts...articulated that technology access for students was for approved *educational or academic use only* [emphasis in original]. This type of policy statement reflects a frame that narrowly defines acceptable technology use for approved classroom activities" (Ahn et al., 2012, para. 43). With regard to social media, Ahn et al. (2012) found that a majority of AUPs made no mention of social media, 34% restricted its use, and 14% banned it. Any teacher should investigate such policies for the district where they currently teach or consider teaching in the future, especially when banning social media would contradict recommendations from the two major teaching groups NCTE and NCSS.

Communication between School and Home

It is now common to see field trip photos, student video projects, and livestreamed graduation ceremonies publicized on Twitter. In much the same way that companies can develop human qualities, so too can school districts when they interact on social media. The Wake County Public School System in North Carolina was recognized for its humorous tweets in response to students' and parents' comments about snow days, in one case calling a student "boo" and in another, calling out another school district because "there's only room for one sassy school district" (Coby, 2013). The following example is a more recent tweet from the same district on the last day of school.

> Wake County Schools @WCPSS · Jun 9
> 💃💃 WE MADE IT! 💃💃 Please excuse us today if you see us dancing around the copy machine. #ohhappyday!
> ↩   ⇄ 22   ♥ 97   ...

Figure 12. School district tweet with emoji (Wake, 2016)

The district displays use of common Twitter strategies such as hashtags, emoji, and all-capital letters to convey extreme emotion (in this case, excitement).

Online Resources for Teachers

From lesson plans, samples of student work, and Pinterest to subreddits, Twitter chats, and online discussion groups such as the NCTE English teachers' discussion board or Jim Burke's (2013) English Companion NING, online resources and aggregating tools for teachers abound. Readwritethink.org offers digital tools such as a stapleless book template and a timeline creator to enhance lessons. Online applications such as *Rory's Story Cubes*, *Writing Prompts*, and *Toontastic* can support students' writing (Karchmer-Klein, 2013). Most of these resources are not new, but the chances to save links for oneself or one's class through a hashtag in Twitter or as a pin on a digital Pinterest board do offer new ways to organize, retrieve, and share such resources.

Digital Writing as New Literacy

In his texts for English teachers, Jim Burke (2013) has highlighted the value of digital writing such as blogs, multimedia texts, wikis, and social media in the modern classroom. Students require knowledge of how to write well in various formats, and online formats are a place where students already write and interact. Some key features of digital writing are its more public nature and the chance for students to explain or enhance their writing with linked images and graphs (Burke, 2013). Multi-modal texts can be more interactive than traditional texts (Greenhow & Gleason, 2012). Furthermore, those who blog "take on the role of both writer and editor, making decisions about the content, layout, and language of their electronic text" (Karchmer-Klein, 2013, p. 313).

Twitter is often called a micro-blogging service due to its character limit and post format; it can be useful for teachers as they teach writing concisely, taking on the persona of a character, or conversing through a back channel during or after in-class discussions (Karchmer-Klein, 2013), Digital writing is, of course, not limited to blogging and hyperlinked essays. Interacting with friends on social media has become a new form of literacy for teens (Davies, 2012; Gleason, 2016). Twitter also crosses over from a site for blogging (or micro-blogging) to one where users can connect with others and communicate in "hearts" or brief comments, which occurs on other social networks such as Facebook. When users learn and follow rules for how to post and comment with emojis, memes, or gifs, they are becoming literate in a new language. Teachers can guide students' literacy skills on social media and in other online interactions.

SUMMARY

- Although there have been warnings about using social media with students, professional organizations for teachers have also called for its use in classrooms.
- Many teachers and school districts use social media to facilitate communication between school and home as well as between students and experts in various fields.

- Most districts have AUPs that guide the ways that teachers and students may use and access social media while at school.
- Social media offers a new form of literacy that students have been developing; teachers may shape and hone these skills through assignments.

EXTENSION ACTIVITIES

1. Locate and follow practicing teachers of similar subjects who teach locally, nationally, and internationally.
2. Search for your own school district and teachers on Twitter and other social media. Describe the online personae that they have crafted.
3. Develop a relationship through Twitter with a teacher who is looking for external respondents for student projects or writing. Offer comments and reflect on how the process and interaction affected you, if at all.
4. Identify a social media-based classroom assignment at the level at which you plan to teach. Complete the assignment and reflect on its effectiveness.
5. Continue your review of the literature related to your own research topic. Determine how these questions have been addressed in other classrooms or related settings.

CHAPTER 6

DEVELOPING AN IDENTITY ON SOCIAL MEDIA

> Just as the actor depends upon stage, fellow-actors, and spectators, to make his entrance, every living thing depends upon a world that solidly appears as the location for its own appearance, on fellow-creatures to play with, and on spectators to acknowledge and recognize its existence.
>
> (Hannah Arendt, as cited in Popova, 2015, para. 9)

We develop personae and identities in our everyday lives with friends and family, and in the classroom with our students. The same is true of our interactions on social media. These interactions can form an individual identity, or they can be one component of a larger constellation of identity construction that includes both online and face-to-face interactions. Gee (2001) notes that one of the ways that identities are formed is within discourses; the thread of Facebook comments or Twitter retweets is a discourse with many similarities—but several key differences—to a face-to-face conversation. As Greenhow and Gleason (2012) noted, "A tweet stream is a constantly evolving, co-constructed conversation" (p. 472). Viewing the profile photograph of a friend or stranger may provide the illusion that one is communicating as might occur in an in-person conversation, but of course the interaction is not as immediate and personal as it is when face-to-face. The nature of social media means that interactions might include more people or extend over a longer period of time, because people can view and participate in the conversation asynchronously.

THE FRAGMENTATION OF ONLINE IDENTITIES

Identity online can be complex and fragmented. Rather than living in one body that walks around and speaks with others, individuals' avatars or profile photographs live and interact in separate spaces. People may even interact anonymously through their avatars, if they are not photos of the people they represent. One glimpse at anonymous (or even full name or Facebook-linked) online comment boards may reveal new lows of human nature. An arm of journalism now even involves reviewing and moderating these kinds of comments (Murtha, 2015). Audiences on comment boards may feel free to fly the flag of their inner racist or attack complete strangers for their gender, sexual orientation, or even their grammar. Teachers may encounter commenters who expound on what they view as the easy lives or simple work of teachers. Scholars attribute this tendency toward negativity in part to the *online disinhibition effect*, which is comprised of six factors: dissociative anonymity,

CHAPTER 6

invisibility, asynchronicity, solipsistic introjection, dissociative imagination, and minimization of authority (Suler, 2004, p. 321). In short, we generally do not see the people with whom we interact online, at least not when they read our messages. These audiences do not see us. Any moderator is likely faceless and involved in only a cursory fashion. This protective feature of online settings allows people to say things that they would not ordinarily say in face-to-face conversation.

Early research on social networking showed that young adult college students considered their audiences when using social media to portray certain personae:

> Both Birnbaum's (2008) sociological dissertation and Saunders's (2008) study reviewed college students' activity on Facebook—Saunders focused specifically on preservice teachers—to determine how they created a persona on the internet. This application of theory extends the originally defined realm of social interactions—face-to-face—into the realm of the internet, and has some important implications for how persona might be formed in the modern day. While Birnbaum found that college students created "fronts" like being social or adventurous and managed others' impressions of them through props, scenery, and gestures in photographs, Saunders found that students were concerned with how to negotiate and reconcile personal and professional identities online, where all of one's friends generally view the same content. (Davis, 2010, p. 28)

As noted above, we behave differently when people can see us as compared to when they can't; it is a "shift to a constellation within self-structure" where there are differing "clusters of affect and cognition" (Suler, 2004, p. 321). The nature of online interactions can be problematic when students and teachers wade into the word of digital communication. An awareness of such issues as the online disinhibition effect will enable those who connect with others online to review and consider the resulting interactions accurately.

SOCIAL MEDIA IDENTITIES: NOT ALWAYS A CHOICE

Identity development and interaction on social media is not always just a choice that new scholars and teachers can decide to make; in many fields, it is not only an unavoidable but also a welcome or expected way of life. Journalists must now consider online commenting (Murtha, 2015). Digital humanities unites computing and the humanities; the origins of this approach have been traced and widely attributed to the work of Roberto Busa (1913–2011) in the 1940's. There are multiple facets to this field, including new forms of data sharing and analysis; use of social media is just one tool that digital humanities scholars may employ. Sharing information on social media is encouraged by scholars such as Jeffrey W. McClurken, professor of history & special assistant to the Provost for teaching, technology, and innovation at the University of Mary Washington, who noted:

It [digital humanities] has tied me into a network of people who are doing—not just similar things—but things that push my thinking and my interests in ways that they wouldn't have otherwise done. And it gives me access to a network of resources that wouldn't have otherwise had. Maybe I could recreate that network without it, but it would take a long time. I [talk about it with the students]; I try to be as transparent as possible about where this fits into my career, about where it has the potential to fit into theirs; I talk about students who have graduated and gone on to get jobs—they are often asked to be the person who runs the Facebook page, or Twitter, or the LinkedIn page for their institution. I've been pretty clear about the connections there and the need to establish a professional identity. These days establishing a professional identity is inherently a digital one…You should be curating your identity. You should be choosing the things that you put out there. (personal communication, January 14, 2016)

McClurken's view of social media adds an element of urgency to the work of those seeking a career, including teachers. As he noted, social media offers a convenient form of networking and allows young scholars to craft the image of themselves—most programs might recommend a professional image, but there are other options—that they want to share.

YOUR VARYING IDENTITIES

Considering your various personae and identities with differing audiences is important because these manifest themselves in both subtle and overt ways when one becomes a teacher. From the clothes one wears to the people with whom one chooses to socialize to the social media presence and events one attends outside of school time, there will always be a kind of trail that students, parents, colleagues, and administrators (prospective or current) may seek out to learn more about the teacher as a person. The following are differing areas to consider as you hone your personal and professional personae and resulting identities.

Your Identity as a Friend

Funny. Serious. Complex. Active. Independent.
 There are thousands of ways that you might describe yourself to your friends, and these descriptors will vary depending on the friend. An individual might wish to present herself as fun-loving, carefree, and well-traveled on social media; she might post pictures with large groups of friends, celebrities, or artfully prepared meals in expensive restaurants. Conversely, one may wish to be seen as quirky and introverted, yet serious, and achieve this by sharing images of her surrounded by her multiple cats and images of coffee cups with the hashtag "#adulting." Reflect on the

CHAPTER 6

kinds of comments and reactions you share with friends on social media. What kinds of posts from your friends do you "like" or "favorite" on Facebook or Twitter? What is the effect or intent of your comments? What are some of the ways that you wish to be seen, and why?

Your Identity as a Family Member

What kind of relationship do you have with your family, and how do you feel about that relationship? Perhaps there is a role that others expect you to play during family events, such as the planner, the joker, or the caretaker. What do you believe about the rights and responsibilities of minors, and how is that guided by your own experience when you were a minor? For example, some might believe that parents should check the social media accounts of their children or befriend them in order to be aware of possible cyberbullying, while others might feel that children should have the freedom to interact with peers without adult oversight. Consider, if you have young siblings or other relatives, what kinds of behavior or images you might warn them about posting online. Are you friends with your parents, siblings, or other relatives on social media? Do you adjust the settings so that they do not see certain kinds of posts or share certain kinds of posts or images only on certain sites where you are not connected with family members? Young adults, including preservice teachers, may maintain Facebook accounts to interact with family and older relatives and interact with friends on Instagram or Twitter more often. What are some of the ways that your family sees you, and how does that align with the person you are on social media?

Your Identity as an Athlete/Artist/Hobbyist/Fan

What sports teams do you follow or play, and why? What are your hobbies, and do you share these with people? What are your motivations for sharing these details of your life? Our interests can become intertwined with our identities whether we wish them to or not. For better or worse, children begin identifying others in this way early in their schooling. In the popular graphic novel *Roller Girl* (Jamieson, 2015), a soon-to-be middle schooler addresses a peer about how others see her at school. She says,

> You know, your 'thing.' What are you known as at school? Like, they call me 'drama girl' because I'm into theatre. What do they call you? (Jamieson, 2015, p. 110)

The resulting conversation leads to the main character's realization that students have a way of referring to each other (such as "child genius" or "horse girl"). This awareness of the ways that interests and skills can become a shortcut to how we view others around us, and in turn how we then view ourselves calls to mind an Eriksonian view of identity formation. Simply put, social media is a place to demonstrate our varied personae, and a place for the audiences from our many settings to understand more about how we view ourselves.

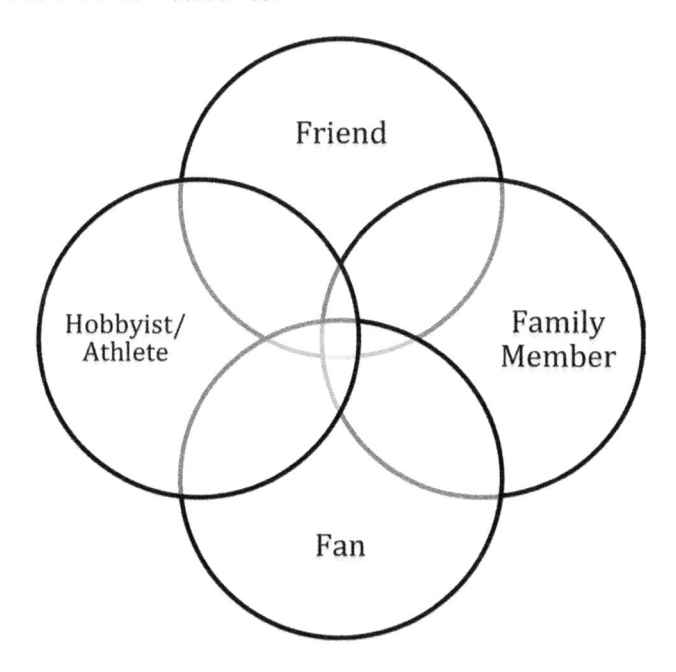

Figure 13. Personal identity factors

The expanded Venn diagram illustrates the ways that various aspects of our identities interact. Imagine that the central point where all of these aspects meet and are made public is the way a person appears to her social media audience.

The following brief case study offers an example of these interactions and how they manifest in the online space: Corrie is a recent college graduate and current student teacher who is a skilled swimmer and the serious captain of her team. She is from a very conservative area in the United States, and has friends with conservative views, while Corrie considers her political stance to be more independent or liberal. While she attended a prestigious university, she follows the college football team of her parents' and sister's *alma mater* (a rival to her own university) because the team is often ranked in the top 10 and she has attended the games and associated tailgate parties (often celebrating enthusiastically) since she was a toddler. When she is not cross training, practicing her sports, or cheering on her football team, she enjoys reading and traveling. She has a large circle of friends (over 900 on Facebook; a quarter of these she considers to be close friends) and a devoted girlfriend of six months. Her parents are unaware of her relationship with a woman because Corrie believes they would disapprove, so she never posts details of this relationship online.

CHAPTER 6

In viewing Figure 13, one can see the various aspects of Corrie's identity that interact or are kept deliberately separate from one another (some even secret, such as her sexuality). In doing so, Corrie is attempting to avoid the problem of "collapsed contexts," where her various audiences might intersect in unexpected or unintended ways (boyd, 2014). While she could choose to select a smaller audience for her various posts to prevent her parents from knowing about her girlfriend, she has decided that it is easier simply never to refer to her girlfriend on Facebook, which Corrie feels creates a strain on both relationships. Many of her friends (but not all of them) are aware that her parents do not know, so using Facebook leads to some anxiety; she feels that she has no choice but to use Facebook, though, because it is one of the key ways that she and her teammates interact before and after games, and she is often tagged in her friends' photos, whether from games, social events, or even family events such as weddings.

Just as Corrie has considered the personae that she presents to her various audiences in online contexts, so too must teachers consider these aspects of their selves so that they can be aware of what information is publicly available and consider how they are perceived by current or future students, colleagues, and administrators. While some may not wish to make any personal details public, we cannot control what others may share about us online. In my early years as a teacher, a student happened to Google my name and located a short story that an undergraduate English professor had shared online with my permission. Reader, in case you weren't aware that students will Google their teachers' names: it turns out that they will. I had long forgotten about the story, but the student had read the fictional story as if it were real, and had been concerned about traumatic themes in the story. After talking with the student and alleviating her concerns, I tracked down that professor after five years and asked him to remove the story, and he did. The internet is a deep, dark hole where countless stories, images, and personal details have been lost, forgotten, and shared with and without others' knowledge or consent.

SUMMARY

- Online identities can be much more fragmented than in-person identities.
- Some careers and settings have expectations for individuals to have a social media or other online presence.
- Considering one's varied personal interests and personae can be a valuable step when determining how and what one will share online.

EXTENSION ACTIVITIES

1. Survey your family and friends about how and why they use social media in their own lives.

2. List the many roles and personae you take on in your everyday life. Analyze how and why these roles differ in various settings.
3. Review the personae of your friends and family on social media. How would you characterize the personae they are presenting, and how does that compare to the in-person roles that you have seen them enact?

CHAPTER 7

DEVELOPING A PROFESSIONAL IDENTITY ON SOCIAL MEDIA

> It's never been more asked of us to show up as only slices of ourselves in different places.
> (Courtney Martin, as cited in Popova, 2016, para. 1)

Just as teachers form complex personal identities, so too do they form professional identities. This occurs face-to-face and in online settings such as on social media. People make choices about the kinds of identities that they will present for a plethora of reasons, but the role identities associated with our chosen career often figure into these choices. Furthermore, we can generally control the emotions that others observe us to have, and we can consider the outcome of these choices on our audiences. As Hannah Arendt (1981) writes,

> …Fear is an emotion indispensable for survival; it indicates danger, and without that warning sense no living thing could last long. The courageous man is not one whose soul lacks this emotion or who can overcome it once and for all, but one who has decided that fear is not what he wants to show. Courage can then become second nature or a habit but not in the sense that fearlessness replaces fear, as though it, too, could become an emotion. Such choices are determined by various factors; many of them are predetermined by the culture into which we are born—they are made because we wish to please others. But there are also choices not inspired by our environment; we may make them because we wish to please ourselves or because we wish to set an example, that is, to persuade others to be pleased with what pleases us. Whatever the motives may be, success and failure in the enterprise of self-presentation depend on the consistency and duration of the image thereby presented to the world. (p. 36)

What might a teacher fear when using social media? Losing a job, perhaps, or alienating a member of one's audience. But no future teacher should allow fear be her sole guide or the primary emotion that others see from her. Pleasing others will be a common feeling for many—although certainly not all—of those who seek to become teachers; however, pleasing oneself and inspiring others can become goals of publicly sharing details about oneself on social media. Setting an example occurs in hundreds of other ways for teachers, and interacting professionally and thoughtfully on social media is one way to set an example for students.

Due to the challenges associated with maintaining personal and professional relationships in a public space, teachers may have more success in both areas if they

CHAPTER 7

keep separate Twitter accounts (one personal, one professional), or use one site for personal connections, and another for professional interactions. An associated (and familiar) recommendation is that teachers should remain vigilant about their posts and avoid offensive content or interactions; in order to do so, teachers can seek out and be aware of several examples and non-examples of these kinds of posts and their effects on audiences. For example, if a PLN discusses the manner in which information has been shared and whether it had a positive or negative effect on the intended audiences, teachers can analyze these kinds of posts to determine if they should emulate or avoid the users' strategies.

PROFESSIONAL USES

Your Identity as a Teacher

What makes a teacher the person he is? Where do his passions lie, and how are these made public to various audiences, including the students? English teachers may love to read and write for their personal enjoyment, but that message can be lost or forgotten in the morass of lesson planning and delivery and the countless other obligations a teacher encounters. A background photo of a well-stocked, multi-colored bookshelf or one's family or pets in a Twitter photo can serve to remind students and others of one's interests and passions.

Viewing the public profiles of one's instructor may be simpler at the college level, where there are fewer restrictions on how and when students use social media, but it is also possible to connect with K-12 students on social media in order to add students to one's audiences. Social media may be used positively and districts may outline practices in their AUPs, although common lore suggests that that teachers should not befriend or follow their students on social media, especially on sites where personal information is most commonly shared, such as Facebook. Fears about sharing an abundance of personal information include that students might misinterpret or judge a teacher's actions or words when they become visible in an online space (as evidenced in the example at the end of the previous chapter). While recommendations to this effect (do not share personal information) abound in the practical literature and common sense might suggest it to be the case, there is little to no research support for this claim.

Some professors or teachers who balk at using social media for courses might cite the memorably named "Creepy Treehouse Effect," where students feel that adults are invading their personal areas and peer groups and therefore it makes the site less inviting to them (Stein, as cited in McBride, 2008). Teachers may have a personal and professional identity on social media, but may choose to stop at interacting with students online. Another solution may be to seek the "middle space," which may mean encouraging students to form groups instead of requiring participation (Couros, as cited in Young, 2008).

Regardless of what elements figure into a teacher's identity, social media offers a unique opportunity to share these with students, parents, colleagues, and administrators, if those groups are a part of one's audience. In this era when instructional time is filled with expectations around high-stakes testing and objectives, presenting personae and public identities on social media can be a way to expand teacher-student interactions, communicate with experts, and form a more complete and multi-faceted identity.

Your Identity as a Student/Alumnae

Preservice teachers or practicing teachers in graduate programs may connect with other students in teacher education courses or, upon graduation, with former classmates from one's time as a college student. Connections may seem most evident among students from within the same programs (a fifth-year M.Ed. in secondary English education, for example), but other connections with student-colleagues across programs (such as secondary and elementary teachers) can lend insight into the vertical nature of education in P-12 settings. For example, blog posts or tweets with images with examples of invented spelling in early grades may inform the background knowledge of English teachers in later grades.

Any teacher has also been a student in the past, and some choose to share their pride for the sports teams or other extracurricular connections to their *alma maters*. Teachers who are completing additional certifications or graduate courses may choose to use social media for course work; this can serve as a model of life-long learning for any students who might view that teacher's social media accounts.

Another point of connection that may be more challenging, but can be facilitated with the help of one's instructor if he or she uses social media, is to locate recent graduates who are now practicing teachers. At a time when accrediting bodies such as the Council for the Accreditation of Educator Preparation (CAEP) seek to track the performance of program graduates, social media can provide a way of maintaining these links when mobility and name changes can make it challenging to keep an accurate and updated list of graduates.

Your Identity as a Scholar/Researcher

A practice that can have a lasting impact is to connect with scholars to conduct research on one's teaching and analyze the impacts of one's practices. Many school districts encourage teachers to conduct research and share their findings with colleagues. Teacher education programs may also encourage action research, often as a capstone project or thesis. In the program where I teach, connecting with scholars online and sharing research ideas on Twitter is a component of courses leading up to the final research presentation. Frequently one of the final tweets

CHAPTER 7

I see of students before graduation features them smiling in front of a presentation screen, immediately before presenting the findings of that research to a group of their professors, mentors, and peers. Sometimes I am the one who takes the photo and tweets it.

In the process of learning about their individual topics for research, preservice and experienced teachers alike can search sites such as Twitter to connect with other teachers and scholars who share their interests. Often authors or organizations will share scholarly articles publicly on Twitter, and educational bloggers will link to useful, related articles. While conducting program-wide research projects, often students will develop interests similar to students who have graduated. Twitter is an ideal platform to connect these students because their resulting interactions are more public than emails. They can bring other scholars into the conversation and share resources easily and concisely through the use of hashtags.

Communication

There are various ways to communicate with others on Twitter and other social media sites. The obvious choice for communication is to post comments or reply to others' comments, but the private messaging feature of sites like Facebook and Twitter allow for more immediate communication than an email. Alerts to a cellular phone are more commonly or automatically enabled on sites like Twitter, so if a student or professor is running late for a meeting, a message sent through Twitter will alert the other party immediately in much the same way as a text message. In this example, privacy is actually increased, because neither party must share details such as a personal cell phone number.

As represented in Figure 13 in the previous chapter, the center of this Venn diagram represents the various aspects of one's professional identities that may be made public to one's social media audiences. In Figure 14 intersection of these three identities is a small fraction, but these three could merge together to form one circle that encompasses all three. Why would one want to keep these areas separate? There are certainly situations where conflicts could occur. One's research areas may not relate closely to one's teaching or even alienate groups of students. Furthermore, when studying for additional certifications or licensure, teachers may not wish to make their goals public to administrators or students, because it may mean that one day they will leave the classroom.

Let us return to the case of Corrie from the previous chapter. In addition to the complexity of her personal life as she negotiated her various identities on Facebook, recall that she was also a student teacher. Several of her teacher education professors began using course hashtags on Twitter to share resources and build connections among students. Her college career center also recommended using LinkedIn as a part of her job search, so, because she wanted to do all that she could to be employed, Corrie created a minimal professional presence on Twitter and LinkedIn for these purposes. While some students from her student teaching position mentioned

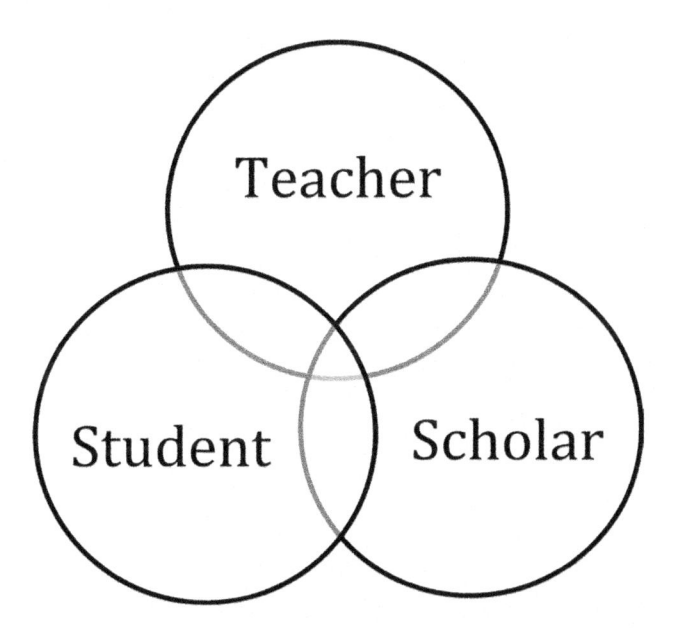

Figure 14. Professional identity interactions

attempting to find her on social media, Corrie wanted to avoid befriending them because she already felt stressed about her various identity negotiations on Facebook. She decided to change her name on Facebook to be only her first and middle names, and to keep her real name on Twitter and LinkedIn, but to only post resources and images that related to her teaching and work as a teacher education student. Further, she carefully considered the images she posted on her personal sites so that they did not overlap with her professional sites.

Could a teacher choose to have a massive seven-way Venn diagram that incorporates all aspects of one's identity in one public space? Of course; this is absolutely possible, but Corrie has opted to maintain clear divisions and boundaries.

This and the previous chapter have considered the many personae that may build over time and in different settings to comprise one's identity. Each of us could describe countless memories, people, and places that have had an impact on the person that we are today. With the advent of social media we have choices to make about which aspects of us we will share with everyone, with many others, with few people, and with no one. Regular and continued reflection on and about these topics is at the heart of how teachers can live in online spaces.

SUMMARY

- Teachers may develop professional identities as teachers, students, and scholars and share those identities on social media.

- Just as with personal identities, considering one's varied professional interests and personae can be a valuable step when determining how and what one will share online.

EXTENSION ACTIVITIES

1. What are your identities? Where and why do you share them?
2. What are your students' identities? How could you facilitate a discussion with them about the roles they play online?
3. Construct your own Venn diagram or other graphic representation with areas for the various aspects of your personal and professional identity.
4. Add labels, color coding, or a legend with symbols to your diagram indicate where the various aspects of your identity appear (or are hidden) in the social media formats you use.

CHAPTER 8

EFFECTS OF YOUR EMERGING IDENTITY

Students can sniff out a manufactured persona in a second, and you will lose their respect and trust. It's OK to be reflectively searching for the best classroom approaches, but putting on an ill-fitting mask makes a teacher look like a desperate superhero wannabe.

(Anderson, 2013, para. 8)

The teacher's process of developing personae and identities can be surprising, exciting, challenging, frustrating—and is, most likely, all of these. Taking on a role and displaying the features and trappings of that role in person and on social media is a harrowing journey for any teacher to navigate. It can feel and appear triumphant and honest or terrible and fake, as evidenced by the above quote.

EFFECTS ON ONESELF

Preservice and Beginning Teachers

A famous and widely viewed TED talk by Amy Cuddy (2014) shares research findings that support the idea that when we "Fake it 'til [we] become it," by holding our bodies in certain ways, that our bodies will then show the physical signs of being more confident, such as lowered cortisol. One of the most stressful times in the journey from student to teacher comes when preservice teachers must present themselves in professional ways during practicum experiences and student teaching, or when a teacher begins a job in a new school.

Practicum experiences and student teaching contribute to preservice teachers' socialization to the profession, and part of that process involves constructing one's role as a teacher based on models of teaching that are both positive and negative (Knowles & Holt-Reynolds, 1991; Lortie, 1975; Wells, 1994; Zeichner & Gore, 1990). However, the process is almost never simple or straightforward.

When beginning teachers experience the process of becoming a teacher, it can lead to confusion or even distaste at the emergence of a new identity that differs from one's previous self (Brown, 2006; Cavanagh & Prescott, 2007) or "reality shock" where the new teacher is surprised and confused that their expectations of pupil behavior and interactions differ considerably from the pupils' actual behavior (Veenman, 1984). Social media can be a complicating factor in this process, as it provides a window into high school or college peers' experiences and career paths, some of which may seem more lucrative or exciting. Articles with extensive comment boards deriding teachers may offend or upset teachers of any experience level. Wells (1984) refers to this shift in roles for new teachers in the title of her paper as "moving to the other side of the desk." Beginning teachers may face several conflicting roles at once or find that their expectations of what kinds of interactions they will have with students do not match reality (Brown, 2006; Flores & Day, 2005; Veenman, 1984; Virta, 2002); for example, one may intend to appear caring and patient, but find herself yelling to a classroom of students or sobbing from exhaustion or frustration during free periods. Time spent with friends in carefree college experiences may give way to early mornings and affordable, conservative teacher clothing. The teacher clothing may even involve humorous, brightly-colored, slightly humiliating seasonal images designed for the students' delight. Furthermore,

> One student in Brown's (2006) study who experienced this identity crisis described hating herself and feeling confusion because she had sabotaged relationships during the course of her identity change, because she felt teachers had to behave in a certain way that differed from her former self. Brown cites Mead's work on construction of the self within social interaction and considers the multiple selves that this student was negotiating, which are

- the person I used to be;
- the person I want to remain;
- the person I hate to be;
- the teacher I fear to be; and
- the teacher I want to be (p. 677, as cited in Davis, 2010).

The emergence and clashing of these five different selves shows the possibility for beginning teacher distress. Because of this identity crisis, practicum and student teaching experiences are critically important to a preservice teacher's development of a teaching persona. These face-to-face challenges are nothing new, but interactions in online spaces during these times are a relatively unconsidered source of additional complications. The sheer number of interactions and the stark differences in expectations of what a teacher is or does can be stressful.

Practicing Teachers

Practicing teachers must also consider their developing and evolving personae and identities. Persona is a key aspect of practicing teachers' identity and interactions (MacDonald, 2004; Wells, 1994). Flores and Day (2005) also name emotions as vital to identity, because teachers must manage emotions in the classroom when they could be distracting or upsetting to students. Teachers are concerned every day with being seen as certain kinds of people, by their students, the students' parents, their colleagues, and administrators. Among the many ways of viewing identity formation are the following four from Gee (2000): Nature, what one was born to be like; Institution, or the position one holds; Discourse, or how one uses language or what is said about the roles we play; and Affinity, or those who belong to our social groups (p. 100). All teachers draw on these sources, and, while all might hold the same position in terms of name ("teacher"), the community in which that teacher lives and works affects how both he and his audiences view and identify him. Similarly, one teacher's understanding of the discourses about what teachers can and should do, wear, or say may differ dramatically from another teacher's ideas.

Identity and Social Media

Added to these concerns about how to form an identity and whether it is appropriate or effective for one's audiences, teachers will enact certain roles on social media. Forming identities on Facebook, Twitter, or similar sites is not without challenges. Users have experienced feelings of superficiality as a result of constructing identities and interacting online. The commonly-used phrase "highlight reel" has emerged in numerous blogs and social media posts to describe what some post; it is a continuous feed of positive images and thoughts that does not give a nuanced view of the actual daily struggles and challenges that are a part of everyday life. As it is impossible to transfer every image and event from our lives to an online forum, it is easy to see why someone might choose only the most flattering images and thoughts to share with others. Some research has found that viewing one's wall on Facebook actually increased self-esteem (Gonzales & Hancock, 2011). Similarly, Valkenburg, Peter, and Schouten (2006) found that positive comments increased self-esteem and negative comments decreased self-esteem.

The other side of the highlight reel is the "echo chamber," a term that writers have used to describe the effect of surrounding oneself with people and ideas that are similar to one's own. Anyone who has observed or engaged in a social media debate about a highly-charged political issue may recall the far-flung cousin or high school friend with differing opinions who is quickly run out the debate or overpowered by many on the opposing side. Teachers who use social media in their personal or professional lives, or both, will experience feelings and have opportunities and considerations that may not be possible or expected in face-to-face interactions.

EFFECTS ON OTHERS

Just as social media interactions can have positive or negative effects on one's identity, so too do these interactions affect people in one's network. Viewing the images, comments, posts, and links that one shares publicly can make others feel joy, amusement, frustration, fury, envy, or indifference. A common complaint is that people share the highlight reel of their lives on social media, so their audiences do not see the negative aspects, struggles, or challenges of real life. Users such as @socalitybarbie on Instagram poke fun at this highlight reel, which has become a kind of clichéd presentation of self that is branded as authentic and creative but becomes mundane in its repetition (Merilli, 2015). The @socalitybarbie account

CHAPTER 8

posts images of a Barbie doll posed in the forest and wearing a knit cap and heavy-framed glasses, with several hashtags such as #liveauthentic and #exploreeverything (Merilli, 2015).

Krasnova et al. (2013) "coined the term 'Facebook envy,' which describes the envy felt after spending time consuming others' personal information on Facebook" (as cited in Tandoc, Ferruci, & Duffy, 2015, p. 141). On a larger scale Bollen, Mao, and Pepe (2011) analyzed millions of tweets and found that "major social, political, cultural, and economic events are correlated with significant, even if delayed fluctuations of public mood levels along a range of different mood dimensions" (p. 453). Conversely, viewing the posts and comments on friends' Facebook or Instagram feeds can reveal the struggles that others encounter and models of how people can respond in thoughtful and positive ways. As I wrote this text, an acquaintance used Facebook to document her husband's appointments, challenges, and move to hospice care as he lived out his final days of a battle with cancer. The outpouring of love and support for her and her family made visible the simple ways that care is expressed for others when it may not come naturally to some. Furthermore, the experiences are in stark juxtaposition to a time when health concerns may have been kept private or even secret; we saw their battle, and we knew that her young daughters understood what was happening and that they had the love of their community. If there is any question, the posts and comment threads will be easily accessible in the future unless the standard practice at Facebook changes; even if it does, some companies will now print collections of posts and comments into bound books for those who are interested.

At this point in time, there is no conclusive research about whether the use of Facebook or other social networks contribute to or exacerbate depression. Some studies have even found that Facebook can lessen depression; however, heavy users of Facebook have higher levels of Facebook envy, and Facebook envy predicts symptoms of depression (Tandoc et al., 2015). The research does not yet reveal a complete picture of what happens when we interact online, but findings such as these indicate that there may be more lasting effects than simply enjoying more connections and communication with friends, family, and others. Any teacher who uses social media in the classroom or for professional development must be aware of the potential effects on herself, her students, and her audiences.

SUMMARY

- Considering and developing an emerging identity can be stressful and have effects on both oneself and one's audiences.
- Some of the ways that social media can lead to frustration, envy, or depression—or feelings of fakeness or superficiality—include the view that people may share only their highlight reel.

EXTENSION ACTIVITIES

1. Conduct a self-study of your social media posts and interactions for the last week or month. How would you describe the ways that others respond to your posts? How would you describe the ways that their reactions affect you?
2. Investigate several of your friends' posts to identify a continuum from positive to negative feelings represented. What are some sources of positive or negative feelings for them, and how does that representation match or conflict with the feelings you recall that person exhibiting in face-to-face interaction?

CHAPTER 9

METHODS OF THE STUDY

RESEARCH QUESTIONS

This study had three guiding questions: first, how do preservice teachers respond when required to use Twitter professionally for a teacher education course? Second, how do these participants use tweets to construct identities, if at all? And third, how do preservice teachers respond to the idea of using Twitter to develop identities based on seeking and sharing professional knowledge?

METHODS: DATA COLLECTION

The mixed-methods data for this study consisted of an electronic survey of preservice teacher candidates who have tweeted for professional purposes in a teacher education course over the last two academic years (2012–2014), and the public tweets and in-depth interviews with four participants who represented a range of responses (positive, neutral, and negative) to the course tweeting requirement. The researcher developed comparative case studies for those interviewed.

Survey

The purpose of the survey was to provide descriptive data about the population for this study and to identify participants for the later phases of the study. The survey was created and distributed through Qualtrics, and included a short section on demographics such as the nature of the respondents' teacher education degree, students' initial response to tweeting requirements, the amount of time spent on Twitter for the course, and students' view of the purpose of their Twitter activity. Open-response questions on the survey were constructed based on the review of literature and focused on students' feelings about identity development in online settings, motivations for their level of involvement, and perceived strengths and weaknesses of the use of Twitter for professional purposes.

The survey was shared with potential participants by an email link sent by instructors who used Twitter in their teacher education courses. These instructors were contacted based on the researcher's knowledge of their use of Twitter as a course requirement. Multiple requests for additional contacts yielded no response, which suggests that use of Twitter for teacher education coursework may not be widespread. Due to the nature of the distribution methods, it is impossible to know how many potential respondents may have seen the survey but decided not to participate. The choice not to share the survey broadly on Twitter, while it may have increased the number respondents, was deliberate; this was to ensure that respondents represented the full range of reactions to the use of Twitter. Those who disliked the requirement or did not participate would not be active on Twitter, and so would not see the survey if it were shared in that way.

The final question of the survey that led to the next phase was a separate electronic form where participants could enter their Twitter handle and rate their overall response to the course Twitter requirement so that a purposive sample could be selected. The electronic form was not linked to unique survey responses so that responses to the survey remained anonymous and were not linked to actual tweets or interview responses.

Case Study Contexts

The four case study participants were white females between the ages of 22–25. Three of the participants were completing a Master of Education program in secondary English, and one was completing the same degree program in PK-12 Art Education. A small mid-Atlantic university served as the site of their shared teacher training, and all of the participants were involved in different local school settings in practicum or student teaching placements. Relevant details about the context of the teacher training program include that the participants all completed at least five 20-hour practicum placements before they began (or would begin) student teaching; they had coursework in instructional design, classroom management, educational law, teaching in their content area, and special education; and they all also had an undergraduate major in their content area.

The courses in which the participants had tweeted were taught by the same instructor but during different semesters in either 2014 or 2015. Courses were either devoted to general secondary and PK-12 instructional methods or action research.

In every case, the tweeting requirement for the participants' courses was a small part of their course grade (equivalent to roughly 2–5%). Other noteworthy aspects related to the tweeting included that the participants used a course hashtag to allow members of the course (or interested members of the public) to filter for only those posts and replies. Additionally, students were invited to create a new, professional Twitter account only for course use if they desired, or they could add to an already-existing personal account. In all four cases, the participants maintained only one Twitter account for both personal and professional tweets. Tweets for the duration of the semester had to remain public so that course members could see and respond to their classmates' posts.

The instructor did not provide explicit instruction in how to construct an identity on Twitter, or even that this was one of the goals behind the use of Twitter. In most cases, the survey for this study was the first time that participants had seen this purpose articulated. Instead, the points that were stressed about Twitter during class were the benefits of connecting with peers and other scholars, sharing resources, and viewing the practices of local and far-flung teachers to broaden the students' experiences of teaching.

Tweets

The second source of data for the study consisted of the public tweets of the four case study participants. The time band for each participant was set as the beginning and end date of the semester in which they had tweeted as a course requirement (either Fall 2014 or Fall 2015). After searching for all tweets within that time band, I copied tweets into a Microsoft Word file for ease of coding. The following provides a more complete picture of the range of Twitter involvement: the participant with the highest number of tweets (11,500 as of October 2015) had 112 followers and 110 following. The participant with the lowest number had seven followers and 47 following, and 40 total tweets (these were ever-changing numbers).

Interviews

Data collection methods for the final phase included in-depth interviews to provide a comprehensive view of each of the four cases (Yin, 2009). The four participants completed an extensive, audio-recorded interview lasting 45 to 60 minutes. Interview questions were grounded in the literature, designed according to Kvale's (1996) seven stages of interviewing, and included the following questions, as well as related prompts:

- How would you describe the person you are with friends and family (your identity)?
- How would you describe the person you are as a teacher education student?
- How would you describe the person you are on social media?
- What are some of the best/worst things about social media (especially Twitter)?
- Why do you think your instructor included the use of Twitter as a class requirement?
- What were some of the benefits of using Twitter for a teacher education course?
- What were some negative aspects of using Twitter for a teacher education course?
- Describe the ways that you used Twitter for your course.
- How does a person form an identity?
- What is your identity on social media?

As noted above, the four participants were selected to provide a purposive sample of a range of responses to the course tweeting requirement. The researcher transcribed these interviews.

CHAPTER 9

METHODS: DATA ANALYSIS

Survey

Survey data were analyzed for descriptive statistics in Microsoft Excel and Qualtrics. The coding of open responses on the survey was aided by NVIVO in order to provide numerical data to support the later findings, such as numbers of instances of each code and interactions between codes. Open responses were simplified to identify common themes; for example, if a participant commented that one of the benefits of Twitter was that it was fast or easy to retrieve information, and another commented that "speed" was a benefit, those two were counted under the same benefit entitled "speed/ease of communication."

Tweet Analysis

The eligible tweets for each participant for the semester in which they tweeted as a course requirement, when collected into a single document, comprised 116 pages of text. The next phase of data analysis involved reviewing and coding the tweets for common themes. The main purpose of the tweet analysis was to triangulate the data by providing an additional, numerical data point (particularly the number of times that participants tweeted to the course hashtag) and examples of actual tweets to enhance the case studies. Actual tweets were adjusted slightly for inclusion in the findings to protect the confidentiality of the participants.

Interviews and Case Studies

The researcher analyzed the data corpus for each case according to Miles and Huberman's (1994) three-step approach to qualitative data analysis. This included data reduction through the use of codes, data display through the use of comparative instances of certain codes, and conclusion drawing/verification. Analysis of tweets included a start list of codes drawn from the research questions and conceptual framework, such as "response to tweeting (positive/negative)," "identity," "persona," "professionalism," "resident/visitor," and "informational/expressive."

The process of analysis was systematic and occurred alongside data collection; case study research calls for this process because early findings can drive later data collection (Yin, 2009). To provide an example of the way in which this process occurred for this study, after reviewing the participants' tweets before the interviews, I was able to refer to actual tweets and potential common themes during the interview to determine if the participants noticed or concurred with these patterns.

It should be noted here that the narratives of the participants were reduced and coded, but the flow of their language was retained to provide a clear sense of each speaker's voice.

Reviewing the data corpus, coding the participants' interviews, and comparing instances of codes to each other to determine common and exceptional themes led to the cross-case analysis of all four participants to determine patterns across the cases and make assertions from the data. Every stage of the analysis process was guided by the conceptual framework. I viewed the data through dual lenses: first, that preservice teachers construct personae based on varied sources to develop identities, and second, that the investment of time and purpose for use are key aspects of how participants used social networking for their teacher education courses.

As noted above, because one potential benefit of Twitter in teacher education has been to make knowledge more accessible and allow individuals to share their own responses to all manner of situations, infused in this analysis process and presentation of data was a high regard for the participants' voices. I did not wish to alter the voices of the participants, and so their comments are not reduced to favor heavy interpretation and analysis. Bruner (1991) indicates that our narratives become our identities, and prior work has shown that the ways that preservice teachers develop and present personae contribute to their identities (Davis, 2010). An aim of the case portraits was to allow space for the participants' voices and thinking about their identity development.

LIMITATIONS

This study has several limitations that must be noted. The survey yielded a small sample that served the sole purpose of describing the population from which the phase two and three participants were drawn. The intent was not to produce generalizable findings, but rather to provide a detailed context description.

Sample

While it has been noted in a previous chapter, it bears repeating that this sample was comprised solely of white females in their early twenties. While the ages of the participants were intentionally limited as an aspect of the population of interest, the uniform genders and ethnicities of the participants is a major limitation. While the range of interests, socio-economic background, and prior experiences varied widely for the four participants, it is important to know how non-white and non-female-identifying students respond to social media and its role in their identity development. Any further studies must investigate the ways that young teachers of color and male or transgendered participants might feel about social media and its role in their creation of personae and professional identities; that was beyond the scope and capabilities of this study.

Additionally, the case study participants represented a range of views about the use of Twitter. The choice to collect data from a purposive sample was intentional.

CHAPTER 9

The intent of the research was not to support one claim about social media and its use in education; however, just as the participants varied widely in their response to the use of Twitter, they also varied in their interest in the research and in the length of their responses. I did not wish to impose a set length on the case portraits because it would risk eliminating key insights, so the quantity of comments differs a great deal from the first to the last participant.

The Nature of Social Media

The frequency of tweets to the course hashtag and the survey responses show that—what may not come as a huge surprise—students completed the tasks because they were required for the course, and not by choice. Making determinations about identity development from required tweets and the participants' comments was a challenge, and more study of these concepts is warranted.

The complicated nature of research involving social media, especially research such as this study that aims to protect the participants' identities while also providing a rich description of their experiences, was a limitation in various ways. For example, when someone chooses to post information on a social network such as Twitter and uses their real name, they have no expectation of privacy: that person is aware that strangers may see or interact with that post in myriad expected and unexpected ways. In contrast, when participating in an IRB-approved qualitative research study such as this one, participants are given pseudonyms to protect their identities so that that they may speak freely and honestly. To combine the public nature of Twitter and the intimate nature of interviews was challenging and problematic. The participants in this study have been given pseudonyms and identifying details have been changed; these include details such as exactly what can be seen in the participants' profile pictures on Twitter and what specific hashtags they used in their posts. Furthermore, I do not share the actual tweets of the participants (but rather similarly-phrased and -themed messages) to protect their identities.

Finally, the ephemeral nature of Twitter meant that numbers of followers and members following, total tweets, and even tweets to the course hashtag (since some could occur after the active semester ended) were constantly changing, so the actual numbers in the context provide a snapshot of the participants' accounts at a given point in time. Any number of events could have occurred in the ensuing time to change the actual quantities, to include something as extreme as a participant deleting her account. To be clear, this did not occur during the analysis process, but it remains a possibility and it highlights the ephemeral nature of social media and the challenges of conducting research in this environment.

EXTENSION ACTIVITIES

1. Return to your research interests and review of the literature to develop your own study. Your study should have the goal of learning more about the effects of some aspect of your teaching.
2. Determine which methods best suit the questions you have about your teaching.
3. Locate and review studies with similar populations and questions.

CHAPTER 10

FINDINGS

The Survey

Twenty-five surveys were returned, which was an approximate return rate of 20%. One respondent did not complete all of the questions, so the total number of responses for certain questions was 24. The exact survey completion percentage is unknown because those who were invited to participate but did not cannot be tracked; this is because some instructors shared the link with their courses on learning management systems such as Canvas or Blackboard. The total population of those who were invited to respond to the survey and who met the requirements of having tweeted for a teacher education course was estimated at about 125. The responses to the key questions of the survey are summarized in this chapter in order to provide a description of the context.

Response to Tweeting

One-half of the respondents used their personal account for class use, and one-half created a separate personal account. The responses to Twitter were mixed; 46% strongly agreed or agreed that they enjoyed using Twitter for a teacher education course, 21% disagreed, and 25% strongly disagreed; the remainder neither agreed nor disagreed. A cross tabulation of time of use and enjoyment appears in the table below.

As see in Table 2, seven respondents reported using Twitter for 10–30 minutes per week and agreed or strongly agreed that they enjoyed using Twitter. Three who used Twitter for 10 or fewer minutes per week agreed that they enjoyed using Twitter. On the negative side, ten respondents who used Twitter for 10 or fewer minutes per week disagreed or strongly disagreed that they enjoyed using Twitter. Only one respondent who used Twitter for 10–30 minutes per week strongly disagreed that he or she enjoyed using Twitter.

Participants identified a wide variety of positive and negative aspects of Twitter, and a third of participants continued using Twitter after the course ended. Some of the reasons provided for why participants believed that their instructor included the use of Twitter as a course requirement included the following (number of multiple responses appear in parentheses):

CHAPTER 10

Table 2. Enjoyment and time of use

I enjoyed using Twitter for one or more teacher education courses.	Approximately how much time did you use Twitter during an average week for your teacher education...				Total
	Less than 10 minutes	10-30 minutes	30-60 minutes	More than 60 minutes	
Strongly Agree	0	2	0	0	2
Agree	3	5	1	0	9
Neither Agree nor Disagree	1	1	0	0	2
Disagree	5	0	0	0	5
Strongly Disagree	5	1	0	0	6
Total	14	9	1	0	24

- Speed/ease of communication and information retrieval (8)
- Create or improve a Personal Learning Network (PLN) (7)
- Become familiar with technology (5)
- Locate resources and other educational material (3)
- Model the use of technology for teaching (2)
- Connect with the professor and classmates (2)
- Make the class more relevant (2)
- Keep discussions going outside of the classroom (1)
- Assess ideas from class (1)
- Create a "professional footprint" (1)
- Share ideas from class with the world (1)

When asked why they used Twitter in the aforementioned ways, 50% noted that they used it because it was required. Additionally, some participants commented that they used Twitter in the following ways:

- To locate, share, and/or read articles and other resources (5)
- To interact with others (5)
- It was convenient (2)

Participants described some of the best aspects of Twitter. These included

- Speed/ease of communication (5)
- Networking, wide reach (3)
- Humorous/entertaining (2)
- Forces writer to be concise (2)
- Seeing what others post (1)
- Informative (1)
- Link sharing (1)
- Public chats (1)

Conversely, some of the worst aspects of Twitter for the participants included

- Limited characters (7)
- Uninteresting/boring material in the feed (5)
- Too public or uncomfortable (4)
- Hard to keep up (3)
- Not intuitive/challenging to learn (2)
- Can lead to jealousy based on others' posts (1)
- Not useful for private or small group discussion (1)
- Instagram is better for image sharing (1)

In response to the question about whether the participants continued using Twitter after the course and the requirement to use Twitter had ended, 56% said no; 34% said yes; and 10% said that they sometimes or occasionally used Twitter.

CHAPTER 10

Identity Construction

The survey concluded with questions about how the respondents thought about identities and how they are formed. The respondents considered first how a person forms identities in general; common responses included

- One's thoughts about herself (6)
- One's actions (5)
- One's interests or experiences (5)
- One's beliefs (3)
- One's connections with others (3)

One respondent's definition was, "gathering your experiences and reflecting on them and then taking action on that reflection."

The respondents then identified how a person can form identities on social media. Some common themes included

- Posting specific things (15)
- Connecting with certain people (3)
- Unsure (2)

Seventy-one percent of respondents commented that one can form an identity through their posts on social media. Several made insightful comments about identity formation on social media. One added that the identities that are formed online can be false, or "an image that people want to be perceived as." Another noted, "it all depends on how they want others to perceive them." One respondent explained

> People choose what they believe to be the best aspects about themselves and present that to the world online through photos, statuses, likes, groups, personal information, friend requests/acceptances, and activity. People are different in real life than they look on social media, but you can still learn something about the way they see themselves and who they might like to be by looking at their social media.

Another respondent described this process as "careful curating of what you post to reflect what you want."

The survey questions about Twitter use in the course appeared before those on identity formation, and respondents could not return to the earlier questions once they had been submitted. Twitter was a place where the participants completed a course requirement by sharing links or conversing with peers. Only one participant connected the requirement with professional identity, which he or she called a "professional footprint." Some aspects that might be considered part of a professional identity, such as "create or improve a PLN," "become familiar with technology," "locate resources and other educational material," "share ideas with the world," and "model the use of technology for teaching" did emerge as outcomes for the participants, but explicit identity construction was not named. Because a survey cannot capture nuances of how, why, and whether identity was formed, the case studies in the following chapter add to this data and these two forms of data are discussed further in Chapter Fifteen.

Twitter for Construction of a Professional Identity

While a majority believed that it was possible to construct an identity on social media, only 17% of the respondents felt that they had created a *professional* identity on Twitter. Far more disagreed: 33% disagreed and 8% strongly disagreed with the statement. Time of use was not a factor that was related to feelings about professional identity: responses were evenly distributed between those who reported using Twitter for less than ten minutes a week and those who reported using Twitter for ten to 30 minutes a week. Only one respondent reported using Twitter for 30–60 minutes a week, and that respondent neither agreed nor disagreed that he or she had developed a professional identity on Twitter. Forty-four percent of respondents noted that they used Twitter because it was required, and more than half did not use Twitter after the course ended.

Practices that increased thoughts of professional identity involved retweeting professional articles, following scholars in the field, and linking to self- or peer-written academic blogs or articles. While only one respondent mentioned keeping a longer-form blog about scholarly topics, that participant also had a positive response to the course tweeting requirement.

Case Studies

The case studies in Chapters Eleven through Fourteen represent the four participants in the final phase of the study, each of whom completed an in-depth interview about their response to Twitter, social media, and identity formation. Each of the case portraits begins with a brief introduction to the participant's tweeting activity for the course in which they tweeted for a course requirement. Each of the case study portraits also includes multiple subheadings that divide the responses into the participant's response to Twitter and social media and their thoughts on forming identities, as well as other key quotes that further organize their comments.

In order to foreground the voices of the participants, the case portraits feature extended quotes from each participant interspersed with my own interpretive and descriptive comments and analysis. Finally, the case portraits have been organized according to the participants' general responses to the use of Twitter for teacher education coursework: the first participant responded most positively, and the last participant responded most negatively.

CHAPTER 10

EXTENSION ACTIVITIES

1. Analyze the data from your study by looking at the results in the form of charts, tables, or graphs.
2. Discuss (in writing or orally, with a colleague) the patterns you see in the data.
3. Plan next steps to explore the data further. What questions would you like to ask your participants now that you have seen their responses?

CHAPTER 11

CASE STUDY ONE

Nora

TWEETS

Nora was the most prolific tweeter of the participants, with more than 12,000 (an ever-increasing number) total tweets during her time on Twitter. During her semester of tweeting as a course requirement, she tweeted a total of 83 times and used the course hashtag ten times (two additional times she used a similar but incorrect course hashtag). Twenty-one additional tweets about course-related activities, such as her responses to assignments, quotes about teaching, and thoughts about leading a church group, did not include the course hashtag. None of her tweets during the focus semester were retweets. In about half of her tweets, she would add emojis or other hashtags, such as #lovethisgroup. Soccer and church events featured prominently in her non-course hashtag tweets. Nora's Twitter profile picture showed her at a church function.

NORA'S STORY

"I'm a Really Open Person"

Nora described feeling as if she had similar ways of presenting herself with all of the different groups of people with which she interacted. She was passionate about playing soccer and attending church.

> I try to be the same person with friends, family, professors, personal relationships, but I also understand that there are differences, maybe in the way you speak to your boss or one of your best friends, so I do take into account those kinds of differences. And it's also the formality. So, you know, 'hey, what's up,' 'what's going on,' vs. 'hi, how are you?' So [the way I talk to people might be] slightly different. For the most part, I try to keep my relationships the same. I'm still open, friendly, warm, and inviting. I'm not hiding anything about myself—I'm a really open person. You can ask me anything, I'll tell you.

> As a student, I try way too hard. I get that a lot. I feel like I work really hard because I know that what I'm doing is going to make a difference. Whether it's in the connections that I make or learning habits or study habits or the material, it's just going to make a difference. There is a financial investment in what I'm putting into this education, but in terms of the education program, I love the

CHAPTER 11

relationships that we've formed and the small community that we have here at this school. It allows you to form connections with your professors. You're not just a face in the crowd in a 300-person seminar, so that's nice.

One exception to Nora's openness occurred in practicum settings when she felt as if she were responding to the expectations of the mentor teacher and the setting. In many cases, she felt as if she were expected to observe quietly from the back of the room instead of becoming actively involved with the students and teacher, which differed from what the instructors in her program encouraged.

During practicum I feel like I'm quiet because there aren't that many opportunities for you to speak. It depends on the environment you're in. Once I had a sixth grade English class, and that was the best environment I ever had. The teacher truly thought of me as a fellow co-worker and that was really beneficial to me. But for other classes at the high school level…they're kind of more at a distance with you and you're off to the side, the observer. So I'm pretty quiet. But when I do have those chances to teach the lessons, it's really fun and engaging and the students…it's different from how they're usually taught. And that's a conversation I've had throughout my experience here…I won't say it's as strong as a *disconnect* between what we're learning here and what we see out in the world, but there is some noticeable difference in what we're bringing to the field. So it's interesting to see how the students respond; it's nice, I think.

While Nora did not use social media as a part of her practicum experiences, she did have positive relationships with teachers and students.

When I've come back, the teacher had asked the students what they thought of me when I wasn't there. They said they loved your lesson, they were really engaged, they asked if you could teach them something else. She even used that lesson for her other classes that I wasn't sitting in on. I just can't speak highly enough of the mentorship I had with that practicum teacher. The others weren't terrible, they do have a standard to uphold, and they take into consideration that I'm on a timeline. Teachers obviously have difficult things going on in their life, difficult situations that they've had to account for while I'm a practicum student. Just not as happy as the situation could be. I still want to have an optimistic view that this will be the best career ever, but…it's both school and also personal things. There have been conversations, like a lot of teachers are leaving after this year because of the extra duties and financial pay. It is good for me, it's not just to teach the kids, but to see what life is going to be like as a teacher. I'm appreciative that I was able to witness those conversations.

The process of reading the situation, responding to what was needed, and shaping her actions based on feedback from mentors in the classroom during practicum experiences—and an awareness of the range of experience as normal and expected—was very similar to the process she undertook on social media under the guidance of her coaches; this process is described in the next section.

Response to Twitter and Social Media: "No, I Gotta Delete That, I Just Can't Have It Out There"

Nora constantly considered what she posted on social media; she had developed this way of thinking when her soccer coaches had also coached her social media interactions.

> On Twitter, it's definitely different [than in my everyday life]. I say that because the moment I got a Twitter account, it was at a point in my life when I started developing a professional background. And so I was following people, whether it was my coaches—that's kind of where it started—the coaches who were a part of my Twitter world were an important part of my life, and so they were the ones, if I was at that point where I wasn't really thinking what I was saying, if I posted something that had maybe a bad word or I'm bantering with somebody else, they'll text me and say, 'you gotta watch that because once you put it out there in the professional world, you can't take it back.' So that's when I started really monitoring myself. It's difficult because I'll post something, and then I'll be like, 'no, I gotta delete that, I just can't have it out there.' And I'll realize it right away. Over the years it's gotten a lot easier to take into account. So now I love posting articles and retweeting inspirational quotes.

> It goes back to a lot of the people that I follow as well. The things I retweet and find interesting I want to share with others. I just see Twitter and other social media networks as ways to connect with people throughout the week, when you may not be able to connect with them in person. So I find resources like that are very useful, which is why you probably see me on Twitter as much as you do.

> I use Twitter much more than Facebook. Facebook over the years has not necessarily lost its purpose, but I find Twitter a lot more the tool that most people are using. I don't know if that's just my age, but that's just my personal opinion...I'll use Facebook just to see what people are doing; [I'll be] checking the news feed and things like that.

> It's not that I use one tool for personal and one for professional: I use both for the same purposes. It's just that I follow more professional people, whether professors or coaches, things like that, on Twitter more. Those are people who are in my network. So that's why I gravitate more towards Twitter.

CHAPTER 11

"You Have to Be Clear It's Not to Be Their Friend or Anything"

Connections, availability, and the opportunity to challenge her thinking were among the benefits of social media that Nora identified. Nora thought that social media could be a way to reach students; however, she added that she would not use it to befriend students, but rather to extend learning.

> The best things about social media are that you can connect with people, experts, friends, family, relatives, people in other countries. And you can connect 24 hours a day. There's that availability option. Social media allows you to have that instantaneous connection. It allows you to share photos, memories, it's like you're there but not, so you feel like a part of their life. It allows you to challenge yourself—it goes into the networks that you follow—but if someone tweets and article or a quote, it's those little things that make you think throughout the day and check in to see what's going on. In terms of the education sector, it allows you to connect with experts in your field, with chats and things like that, but also a forum to extend learning beyond the classroom. And that's something that I've really been thinking about a lot this semester, in one of my education classes with technology.

> Twitter is a component to that for us as teachers, but it also makes me think, 'how can I apply this to the education where I'll be teaching students.' So I think of that in terms of group chats, or everyone has a Twitter account, or Facebook groups…Of course you'd create a professional account. You gotta make sure you have a professional environment. You have to be clear it's not to be their friend or anything, so those lines are distinct. Social media just provides that opportunity to extend learning beyond the classroom, so that's another beneficial aspect.

The most negative aspect of social media for Nora was the feeling of permanence; although she had noted that she often thinks carefully about what she posts, this attention to that as a negative aspect suggested that it was not always a positive experience to have to carefully monitor her statements.

> Bad things…that was the number one thing, I still struggle with that, I forget that *this is permanent*, it's important to make sure students understand that as well. I wrote a lesson for self-image and identity in relation to technology and it just shows that what you say matters and you have to consider your audience. That's a downfall, if people can't use it in a way that they're cognizant of who's reading it now and in the future. The instantaneous quality of social media…you can say something in three seconds, without even thinking, just type it real quick, and then, 'oh, no, I didn't want to say that.'

"Why Not Meet Them Halfway and Integrate Ourselves and See What They're Doing and Saying in the World?"

Nora happened to take multiple classes that used Twitter for various purposes, and she acknowledged that she could tell that its use in one of the two courses that the four case study participants had in common was to develop a professional identity. Beyond that, she appreciated the chance to learn about a tool that might be useful as she taught her own students in the future.

> I've used it for a lot of classes. Most have been education based. We tried to use it for one education class, and it was more if you participate in an online chat and show proof you can get credit. That was attempted, but nobody really followed up with it. I tried…[it was like] 15 minutes, I'll see what's going on…I can't remember what the specific chat was. I don't think we used it [in English classes] but that's a tool that could be used, not just for education, but across content classes.

> I think we did it in class to develop our professional identity, and showing that it can be used as a tool outside of the classroom. Here in education it's not just a matter of saying, 'hey, use this with your students,' but experiencing it and seeing for yourself if it'll work. I'd like to think that all students have Facebooks and Twitters and stuff, but not all students do. This semester as I'm reading other peers' comments, it is a requirement for another course, and some are saying, 'wow, I'm surprised this is so new to me, it's such a useful tool.' We're not all familiar with the tools, but here are some ways you can use them. Just the popularity in general, we're using it and students are using these technologies, so why not meet them halfway and integrate ourselves and see what they're doing and saying in the world and what they're passionate about. The personal/professional lines are distinct…I'm following three or four professors, and others are popping up that I should follow and I'm like, 'oooh!' I feel like students would want to reach out.

Nora acknowledged that her extensive prior experience with Twitter made the use of it for a class intuitive and simple. She described times when having a broader audience made her more excited about course work or motivated her to share resources.

> Most people who already use it might not have trouble doing it. Number one…in a class where it's mandatory, we have a checklist with things like a blog post, and you have to submit it within a two-week period and there's a deadline. Every now and then I'll think of something witty to say. I mean, that's the reason why I would post something, I don't want to post something generic, just a definition, but if it's the weekend before I'll have to set parameters, this is due, have to post it. One example I created a Glogster and

CHAPTER 11

> I was fired up and I did the screenshot for that. I sent the screenshot of a collage—it was cool.

> We do have so much to do for that class, but I don't know if I'd do it as much. She made us create the account, and I'd probably use it now and then…I created a separate account because she was stressing the professional image you want to have in the world, but I feel like my Twitter is already professional. After this class I don't think I'll use that account again. Because I've tried to keep that one professional. Especially if something is truly alarming, I hope that one of my professors would be like, 'alright, Nora, let's talk about this,'" I'm aware of the audience that I'm in contact with this. I do have stuff about religion, but it goes back to the fact that that's a big part of my identity. It's been clear since the beginning of my courses that you can't promote your religion to anybody, but I feel like your social media allows you to have that be a part of your identity. You're not in your job making those posts, it's not connected and slightly different. I do understand that that could be a professional roadblock in the future. I don't see myself locking it down [when I go on the job market]… if I lock it down or hide the account, they might be like, 'well, what are you really hiding?'

Forming an Identity: "I'll Show Challenges…This Is Who I Am, and It's Okay for You to See That"

Nora developed a comprehensive identity on Twitter over the course of several years. Various aspects of her life were integrated into the online space, and she felt comfortable with her public identity as one that combined both personal and professional details in one space. This was in stark contrast to some of her peers, who she felt were not considering their audiences carefully.

> It goes to your surroundings and the experiences that occur…also, the personal factors in terms of the people that you're in relationships with, so that's how I see one constructing an identity. On social media it's relatively the same, the people that you're in contact with. I can't say definitively that this is why my friends post what they do, but people who are posting pictures of them getting crazy on a Saturday or using certain lyrics…that may be derogatory, they are in contact with certain groups of people where that is acceptable. And the followers in their network are okay with that and it's not seen as something that they don't approve of. So for me, I'm taking into account the people I follow and who follow me, reading what they're saying, and that forms who I am and what I value and that is shown through the posts that I make.

When considering why one might share only one side of themselves on social media, Nora noted that it can be not just about a lack of consideration about professionalism, but that it can due to their desire to present a "perfect" self. Nora chose to share other sides of herself, and had been thinking about how she presented herself online for several years.

> I'm not a person who will totally shut down the personal from the professional; I'll show challenges. It goes down to this is who I am and it's okay for you to see that. I try to be as open as I can. We're not all perfect. I used to feel that social media was my one true outlet. In creative writing I would term it as my invisible audience, they're out there, someone's listening, but I feel like I'm being heard and I can express my thoughts and feelings. It shapes who you are. We had to create our own domain for the other course, and I was like, 'I hate this,' but now I like it because people are replying to me, even though they're probably spam people. Half are thoughtful, so now I'm like, 'oh, I have an audience, maybe I'll just spice up my profile'… I [recall that I] had a blog in like eighth grade and it was this really girly pink background [and I have changed since then].

Due to her awareness of her varied—and active—audiences, Nora described things that she knew she shouldn't share on social media. Her church was on three different social media platforms, and she often used hashtags to communicate with members of her church and to share images.

> I can't show things like swearing. That's been a roadblock. You're angry at someone but you don't want to be seen as that like 20 minutes later or the next day—at least I don't want to be viewed as that kind of person. We all have those thoughts of someone who might be dressed different from you, but I just don't want to be viewed that way. Swearing, criticism, anything that could ruin professional relationships. There are people who are following me who are great resources…I don't want to make them think that I am different than I say that I am. Controversial discussions are kind of interesting. I've learned the hard way over the last few years. Religion is one thing I try to be aware of when it crosses boundaries…I might reply to one person as a thoughtful message, and the feedback keeps coming and coming. I posted it, the lesson was learned. When Christians are seen as the dominating opinion [I try not to get into debates with people].

> The things I share—I tweeted about church events, and I did think about how the people that are following me will see that. I'm not forcing my views on others. Religion is a strong thing that appears on my profiles. My church is on Facebook, Instagram, and Twitter. And soccer. I have to be careful, because there are professional coaches that I follow.

Even though Nora had developed a careful process for thinking about her posts and making them for courses and for her own personal interests, she did wish for more of an outlet for personal or honest writing. She had experimented with riskier online behavior in her youth; Nora described how she and a friend had developed a secret

CHAPTER 11

Twitter account, where they only just followed each other and did not use their or anyone else's real names. She also referred to "subtweets," which she described as a tweet that would have a subtext that others might understand, but that would not get her into trouble.

> I did once have a secret Twitter account with a friend in school...It just provided an outlet for us and we felt like we could let our walls down. I wish that social media sometimes allowed us to. For some people it is, who aren't thinking about the professional things, but it's been a huge strain for me on a lot of occasions. I'd want to make a tweet, but I couldn't let my whole guard down. It was like subtweets where everyone knows what you mean...right after a class or something.

EXTENSION ACTIVITIES

1. Discuss the process by which Nora developed her professional Twitter presence. What were some contributing factors that she identified in her narrative?
2. To what do you attribute the prolific nature of Nora's tweets? Why do some people participate by posting or commenting on social media much more than others?
3. Construct a matrix or other graphic representation to identify the various contributing factors to how you develop your own social media presence.

CHAPTER 12

CASE STUDY TWO

Marina

TWEETS

During her semester of tweeting as a course requirement, Marina tweeted a total of 29 times. Twenty-one of those tweets included the course hashtag and four to the course hashtag were retweets. Her tweets were evenly divided between course-specific activities such as sharing the name of a scholar in her area of interest and posting humorous images, gifs, or thoughts about student teaching such as "I'm so old now," or "gotta love using popular songs to teach grammar." In her black and white profile picture, she smiled at the camera and wore a scarf, blouse, and dress pants.

MARINA'S STORY

"It Is Kind of Like You're Stuck in One Persona That May Not Be Your Own Persona"

Marina identified herself as introverted, but comfortable around her friends and family. As a college student she sometimes felt nervous about her level of preparation around other students who would complete additional, unrequired work:

> My family pretty much knows everything about me, we all just have a very normal relationship—even my close friends. Select friends, the ones I've been living with for the past five years. Two friends.

> As a student I'm very organized. I do think I'm a little less open as a student in the classroom. I guess I'm an introvert. I generally have thoughts and feelings; I just never express them in the classroom. Which is weird 'cause I'm very open with my family and as a teacher in the classroom. I think it's…being an English major in classroom settings, English majors are always on top of everything. They always seem to have all this free time to read all these books, and they're like, "let me relate it to this book, and this century," and I'm like, I only have time to read what we're reading in class. So I think of that pure intimidation, but no so much this fifth year because we're all very close and very open, but in previous courses.

CHAPTER 12

> In practicum, I'm not like that, I'm—I don't want to say reserved, because I don't sit back and not let anything happen—but more reserved in owning my own teaching style because I am new, and you're in someone else's classroom, you don't want to take over and take control. But I am so willing to sit there and talk with students and do any extra work I need to and be open with my mentor teacher about how I'm feeling and that kind of thing. But it is kind of like you're stuck in one persona that may not be your own persona. The practica that were better than others had more involvement in the classroom. There were some teachers who were totally fine with me just sitting there observing and there were some who would push me to participate in class even if it was just passing out papers or calling roll or making that connection with students.

Response to Twitter and Social Media: "There's Something Very Stressful about It"

Marina referenced earlier discussions in a different course that had identified legal dangers that can occur on social media. Her parents were one of the main reasons that she kept a Facebook account, and she described her dad's interest in making connections on the site.

> I'm a very paranoid person on social media now. Because in an earlier class—I'm super paranoid that a parent or administration or my teacher is going to see something bad or inappropriate or…everything's private, and if it's not private it's for an education class where I'm professional and in teaching mode. I used to have two but now I only use the professional one. I use Facebook, Instagram, Snapchat, Pinterest, but most of them are just pictures, I never really post updates. I very rarely share articles on there. It's mainly just to keep in touch with family and if I go on a trip, I'll post pictures from that. My parents are on Facebook…that's another reason I keep it, my dad is all about Facebook messaging. He's very into reconnecting with old high school friends…even if they weren't friends, he'll just be like, 'hey, this person from school said this.'

> Some of the best things are using it like my dad uses it, for reconnecting with old high school friends, or the way we are supposed to use it in class. Our cohort is very close; we all keep up to date with each other on Twitter for those who do post. I think the worst is that it's social, so if you post one wrong thing, even if you delete it, it's on someone else's, or it's on Google, or it's over here, so that constant access from anyone who can see anything—there's something very stressful about it. In undergrad I used it for a few courses, and I wasn't a huge fan because I'd have to spend a good amount of time looking for scholars or looking for PLC communities, and it just got stressful—because when you think of social media, in this generation, people use it to procrastinate, they use it as a distraction, so it was stressful to combine.

CASE STUDY TWO

After attempting to toggle between two accounts, one personal and one professional, Marina found it confusing and decided to stick to just one professional Twitter account. She did maintain personal accounts on Facebook and Instagram.

> I think in the first class I used it I used my personal account, and it was just irritating to look for scholars and communities when all I really wanted to do was tweet about how my day was stressful. I [thought about non-teacher friends not wanting to see it] and that's when I made my professional one. I thought, one, that they're going to think, 'what the heck is this girl talking about?' or two, 'why is she—this isn't important, I don't care about this. I don't want to see this.' So I made the professional one. And I also made the professional one because I didn't want to take my personal one off of being private. That was also another irritation: juggling two separate ones. Because I'd obviously stay on my personal one on my phone, and then I'd realize the day an assignment was due that I'd have to go check the other one.

Marina noted the benefits of Twitter, including that it enabled connections with professionals. In response to the information that it was popular at English teacher conferences, she attributed that to her field's variety of interpretations and teaching methods.

> I think the reason we used it was to show the different workings of social media, to show that it's not just used for the procrastination or distraction, that it can be used to show different people in your field, or different ideas about your field. And it can also be used to ask for help. If you tweet at someone they may or may not respond, but you still have that connection, like for this last assignment we had to find a person pertaining to our research and I could follow her and see all of her tweets, whereas before Twitter I'd have to track her down or go to a seminar that she's holding... it's just a more direct connection. [So many English teachers might use it at conferences] because there's so much variation in English instruction. Not that there isn't for other courses, but in English there are so many interpretations and modes of addressing one novel. You have 20 million main ideas that you can do.

> When we used it for a technology course, it didn't really expand on using Twitter, it was just to find an article or a specific person to follow, whereas I would have liked to find out about how to incorporate it into classes, and what kinds of assignments you can use Twitter for. And I used it for another non-education class, which was a lot more open and—not less professional, but you could say whatever you wanted, it wasn't that teacher mentality of being super professional and making sure no one's going to see it. It was talking about raunchy things over here, and murder, and femme fatales over here, so that one was more interactive.

CHAPTER 12

Forming an Identity: "If You Use It in a Positive Way, It Can Show Your Influence on Others…"

> We create identities through our choices and how we act on those choices. You can think one way but if you act a different way it's kind of forming who you are—like, 'I know this is wrong' or 'I know this is right,' but if you still act one way, or don't act a certain way, it talks about who you are as a person. So it's those choices and actions. I think the choices come from the support you've had, who's raised you, what kind of community you were raised in. Which is funny because that's part of the reason I was moved from one school to another when I was growing up, because my mom did not like the personality of the people I was hanging out with. She says it all started when I wanted to be called by a nickname and she was like, 'nope, we're going.'

Marina contrasted identity formation in person as compared to on social media. She noted that people can have different identities because it is more public and not face-to-face communication.

> On social media—I'll use an example—some people do the passive-aggressive tweets or posts at someone, where they're obviously talking about specific person, but they don't say who that person is. It's more anonymous; they don't feel like they have to say the person's name. So it can make them feel better about the situation whereas in person, would you ever actually say that to a person's face? It brings out a different identity. Maybe sometimes negative. But I also think if you use it in a positive way, it can show your influence on others or your ability to not be so negative. It's kind of an alternate identity that you don't see in person. Which I guess for some people could be the real—who they really are, versus who they present themselves to be. I feel like it's more a way for people to hide.

When considering how she created her professional identity, Marina thought of herself as careful, even on her personal account on Facebook, where she deleted images of drinking even though she was legally able to drink and was doing so with her family as part of a cultural tour.

> I think people think I'm…it depends: on my professional Twitter, I'm the teacher, the one who's not going to post anything inappropriate. You'll see professional accounts with it [that I follow], you'll see reflections on student teaching, never anything bad. Which you'll never see that on any of my other social media, either. There, on Facebook or somewhere, you'll see me having a social life. I got rid of Yik Yak because Yik Yak here just drove me insane. It was constant arguments about rights and people being degrading and insulting. It did keep me up to date on things that were happening on campus, I will say that. If I didn't know what was going on, I could just pull up that

app and know. I don't swear or anything, even like a glass of wine on the table I won't post it anywhere. When we went to Scotland with my family for a whiskey tour, I didn't even want to post those pictures anywhere.

EXTENSION ACTIVITIES

1. List some of the best and worst aspects of social media that Marina identifies in her narrative. Do you agree with this assessment? Why or why not?
2. Describe in writing how you believe you and other individuals construct identities, both online and face-to-face.

CHAPTER 13

CASE STUDY THREE

Callie

TWEETS

Callie had the lowest number of tweets during the semester when she tweeted for a course, but she did not have the most negative response. She tweeted seven total times during the semester and used the course hashtag only four times. Those four times were course-specific assignments such as sharing possible research questions. She did not retweet to the course hashtag. Her profile picture depicted her playing volleyball.

CALLIE'S STORY

"I Think That Actions Speak Louder Than Words, Especially in the Art Classroom"

Callie described presenting herself in a wide range of ways, from silly and easygoing to hard-working and quiet. She was experiencing some stress at an increase in family responsibilities during the semester in which she took the course with a tweeting requirement. She noted:

> With my family I am silly and easy going. I easily take on larger responsibilities than I probably need to. I am trustworthy and a hard worker and I don't like to ask for help until I am desperate.

> As a student I am quiet but easy going. It takes a lot of trust for me to open up to a new person and I often don't until I really get to know them. I can work in a team setting but I thrive working on my own time and independently.

> As a student teacher I am kind, engaged, and too quiet, but I am working on that. I show the students that I am genuinely interested in what they have to say but keep them on topic by redirecting. I think that actions speak louder than words, especially in the art classroom. Modeling is key.

Response to Twitter and Social Media: "I Am...a Visual Person, So I Prefer Not to Read Peoples' Gossip Very Often"

The actual capabilities of the different social media sites contributed to Callie's response to and use of them. She favored sharing images on Instagram instead of

CHAPTER 13

text posts on Twitter or Facebook because she was an art student and preferred visual images to text. One of her tweets to the course hashtag featured artwork about teaching that she had found online. She said:

> I am very filtered on social media. I do not post much especially on Facebook or Twitter, but on Instagram I post pictures frequently. I feel that on Instagram I have a smaller audience who sees my feed and I am also a very visual person so I prefer not to read people's gossip very often. I think pictures are the most honest form of social media, although editing and such can go on you have to have something to begin with.

In addition to viewing and sharing images, Callie used Facebook for both personal and professional connections, but she found it overwhelming at times. Twitter remained a place where she completed course requirements, but she did not describe connecting with others as she did on Facebook:

> Facebook is all about connections for me, I connect with old teachers, old friends, groups such as other Boston terrier lovers and fellow art teachers and almost everyone has a Facebook, so it becomes very overwhelming. On Instagram I connect with artists, friends, public sites such as local towns, as well as particular brands. Instagram is by far my favorite for my love of adventure and imagery. I am pretty sure I created a Twitter account just because everyone else had one and I didn't use it very much until I needed to for education classes.

"If You Aren't Checking That Hashtag... You Are Missing the Benefits of the Connections"

Callie's response to tweeting for her course was mixed, and she did not tweet or check Twitter more than a handful of times during the semester. One frustrating aspect of Twitter for Callie was its limit on text characters, which she disliked having to edit; she noted:

> On Twitter the amount of words overwhelms me as I scroll through, but at the same time the limit of characters when posting also bothers me. I know this is a contradicting statement, but it really frustrates me when I am trying to add a hashtag or a tag at the end of a tweet and it doesn't fit so I have to go back and edit my tweet. I feel there should be a little wiggle room.

Another frustration was related to how people presented themselves online, which Callie linked to an effect on their audiences when they felt that their lives didn't measure up:

> Social media in general gives people the opportunity to create a false image about themselves simply by cropping certain images or stretching the truth. This opportunity cringes me especially in this day and age where people have

all these mental illnesses that cause them to endanger themselves. People will buy into these stellar lives and wonder why theirs aren't so amazing every day as well; and [that will] cause the viewer to get down on their self-esteem.

While acknowledging that there were challenges; Callie also found some benefits to the use of Twitter, among them that it was simple and accessible, but she added that it required time and attention to realize the full benefit.

I have used Twitter to connect with my fellow classmates and professors sharing research question ideas and various articles regarding education. Twitter allows for the professional connection of other educators as well as other student teachers. Our class hashtag allows all the student teachers to connect with various questions or ideas while student teaching.

My class keeps in touch as we are student teaching; we have the possibility to share what we are working on or how we are liking student teaching, as well as our progress on our research proposals. Social media is used on a daily basis for most people and is extremely accessible. When you look on the App Store on an iPhone, both Facebook and Instagram are in the top five free apps in the charts. Using Twitter as a class requirement isn't asking us to adapt to a new system or a way of doing things most of us already used it. The best has been the connections! I've learned through making my own connections that they can make things you never thought possible, possible. However, our feed is filtered through a hashtag and if you aren't checking that hashtag as a student, then you are missing the benefits of the connections that you could be making.

Even though she recognized the benefits, Callie added that she used Twitter for the course only because it was required. She also stated that she might continue using it if she were aware of useful hashtags.

I wouldn't have initiated the use of Twitter if it weren't for my professor, but the hashtag was predetermined and the entire class bought in to the movement both as a requirement and to keep connected. Not all posts in the feed are for an assignment.

I currently do still use it, but it is still a requirement. I think I would use it in these ways if I knew of a hashtag that all art teachers were using. I could connect with these other art teachers and share project ideas similar to the "Art teacher" group my mentor teacher invited me to join on Facebook.

"We Need to...Start a Movement to Mirror How Students Are Learning Outside of School"

Callie articulated one of the key purposes of the Twitter assignment, which was to be aware of how to use social media because it affects the lives of students.

CHAPTER 13

> We are using social media because the students we will be teaching are growing up with social media as a norm and these students are signing on to these accounts at a young age. As teachers we need to know the most current technology that our students will be using and start a movement to mirror how students are learning outside of school—through social media and the internet in general.

Forming an Identity: "Sometimes I Just Want to Tell People to Cut the Shenanigans and Show Us Your Real Life"

Honesty in social media was important to Callie. She resisted the idea that identities can be formed on social media, because people could be presenting false images or details about themselves. She disliked the constant stream of communication but added that if people were honest, that they could build a professional identity through connections and careful posts.

> People form identities in person. A person's identity is judged on their attitude and moral character. On social media they can form an honest identity or a false identity. Unfortunately, it could be really easy to form a false identity on social media through the use of cropping images to make your life seem more brilliant and magical than it actually is or they can stretch the truth. Some people post about every little thing they do such as what they had for dinner that night; I understand if it was some special event but the constant update of every part of your life can be a little absurd. Forming an honest professional identity can be done by making thought-out posts or tweets as well as the relevant connections made.

> My Twitter account for the past two or so years has been strictly for educational use. My Instagram shows the world my adventurous life, mostly art I've made and pictures of my dog, but it's the connections I make on Instagram with other artists that are the professional connections. I rarely post things on Facebook, but certain aspects of it I use for professional such as the "Art teachers" private group to find ideas of various aspects of art teacher life in the classroom.

When considering what she chose not to share on social media, Callie described thinking carefully about what would remain permanent. She valued honesty but also acknowledged that people can misinterpret or even use information against others in the future.

> [In terms of barriers,] I am very critical about the things I put out there on social media. Sometimes I just want to tell people to cut the shenanigans and show us your real life, but I don't, because that would be the image I then made of myself. It's a lot different to tell someone that in person; it's not recorded as a permanent document as it is on social media. What we say on social media

can always be back-tracked and used against you for whatever reason. Some social media discussions can get very heated and I've known of some people who like to make comments heated on purpose just to see the way people react. This technology is still pretty new to us and if teachers can bring it in their classrooms to then teach our students how to responsibly and professionally use it, that would be a huge step in the right direction.

EXTENSION ACTIVITIES

1. Compare and contrast Callie's narrative with Nora's. How do their experiences and beliefs differ, and why do you believe this might be the case?
2. What might have increased Callie's professional interactions on social media during the semester?
3. Construct a case study of yourself to illustrate (in words, images, or both) the ways that you use technology, both in your personal life and your professional life.

CHAPTER 14

CASE STUDY FOUR

Kate

TWEETS

Kate responded most negatively to the course tweeting requirement. She tweeted a total of 13 times during the focus semester. The course hashtag appeared in 11 of her tweets, but in one additional tweet, she used a similar but incorrect hashtag. Five of Kate's course tweets featured memes or other humorous images related to her thoughts about teaching. One of her course tweets was a retweet. In her profile picture, she smiled as she wore a graduation cap and gown.

KATE'S STORY

"I'm Just Kind of a Goofball"

Kate reflected on her easygoing, silly nature in person, and how that carried over into what she chose to post on social media. She stated,

> I feel like I'm really relaxed. I'm just kind of a goofball. I'm laid back, but there are some things I can get irrationally upset about. Like I don't get mad easily, but then if I do, I kind of go off really easily and then five minutes later I'm fine. So I think I'm pretty relaxed. A strong, foofy person. On social media I'm the same way. I'm not going to post anything unless it's going to be a positive impact on someone's day, like make them laugh. I'm never going to post something to be like, I hope someone cries when they read that. That just not who I am.

> As a teacher education student, I'm surprising because my personality really doesn't change. I'm not really good at being fake or faking it, so even in the classroom, I'm pretty goofy, just like how I am in my classes. There are times when the voice will come out and there's no way to not have that, and I could be like my goofy, ADD self but still get stuff done and I have to remind my students of that. Like you can still have a personality and get your work done. And I think I'm the same way when it comes to me being a student. I can be kind of goofy, but I'm still always getting it done.

CHAPTER 14

"I Felt It Best to Try My Hardest to Suppress Who I Am"

The idea of and resistance to falseness coupled with a perceived need for it arose often for Kate; she had experienced similar feelings in face-to-face interactions as well.

> In a couple of my practicum placements I felt [as if I had to be fake and] uncomfortable. Not because of the person they were, but because I knew my opinion wasn't needed in those situations and my personality, the way that I am, wasn't really something that they were used to dealing with when it comes to people who want to be teachers.

Kate described her approach as one where she avoided any problematic interactions in the future in order to continue in the program:

> So I just tried to stay quiet, which was hard, and which got me kind of mediocre eval[uation]s in one of those classes. Just a little bit of trouble a couple of years ago. I said something to a student that they teacher heard and thought I was undermining her authority as a teacher, which I didn't think was so detrimental. But then I saw it from her side and I apologized and explained my side and she understood. So…that was a situation where I felt it best to try my hardest to suppress who I am.

Response to Twitter and Social Media: "It Allows All Types of Idiots On"

Social media was a mode of personal—not professional—communication with friends and family for Kate, and she used it sparingly. She noted that she used Facebook because her family was a part of her audience on the site.

> In my personal life, I don't get on Twitter. Actually today, when you messaged me (for the interview) that was the first time I've ever used the Twitter messaging. It's genius…it kind of works the same way that Facebook messaging does. I'll use that—I guess I'm a little behind the times, but I only use Facebook; I'm not on Instagram. And I'll post pictures and photos and stuff like that to my Facebook because that's the only thing my family has too.

Communication and the potential for positive distraction were positive aspects of social media, but access in her rural hometown was limited. She saw some clear negatives along with feelings of falseness, among them that it was a place for people to share unimportant or unwelcome views to a wide audience.

> The best things about social media are connecting with family—well, not my immediate family, not like Mom and Dad, they don't have Facebook. Well, my mom does, but she never has the internet to get onto it. Other best parts are that it's a distraction, which is nice. You can use the messaging to talk to people you haven't seen in a while. You can keep up with people…even if it makes

you feel sad because you're like, 'good, you're all getting engaged and moving on with your lives, and I'm over here like, I had chicken salad for dinner.'

But the worst is just scrolling through and realizing why you hate so many people. You're just scrolling through like, ugh, looking at friends and thinking, 'why are you so ignorant?' Social media doesn't block ignorance; it allows all types of idiots on, sharing opinions and whatever.

Kate encountered some difficulties when tweeting for class, which showed in her use of an incorrect hashtag. She also acknowledged that the size of her network was not large and therefore affected the content she saw. She said:

[When using it for class,] I tried to tweet, but I'd be hashtaggin' the wrong stuff, and I would have liked Twitter a lot if I were like famous and had a bunch of followers. I liked it for a minute and then I was like, this is irrelevant. It just didn't matter and I wasn't seeing enough. I may not have tried hard enough to follow the right people, but I could never figure it out. It was hard for me to get used to. But I used it, in classes [when it was required].

While it is generally recommended in teacher education programs that teachers avoid connecting with students on social media that includes personal content for either party, Kate actually described seeking students out to learn more about them; however, she felt conflicted about her searches. She also disliked the idea of having any other social media accounts, because she feared for her current and future job security.

My students tried to find me on Facebook, but honestly the reverse of that is that I've tried to find them on Facebook, but it felt weird. I would look them up to learn their names and see if I could make a connection with something about them, but then I felt like a creeper. I know darn well they're definitely looking me up, because I've had students be like, 'hey, what's your first name, we want to see what you were like in high school.' It's a terrible idea…but then I haven't had them say that they've found me and then they would want to know my Instagram and I'm like, there's no chance in heck that I'm going to get an Instagram. There's no need. That's just like another way that I'm just going to potentially get fired.

One way that Kate handled her concerns about her students or future administrators finding her on Twitter was to mask her real name.

On Twitter I made it really hard to find me, but all the things that I'd have to say on Twitter are just funny or witty or my sarcasm that might not be school-appropriate. I wouldn't want students or administration to find things, so I just try to avoid it in general because I don't want something that I say to come back and bite me. I'd probably just do that just living my life, I don't need social media to help me. [That might be related to my earlier practicum

CHAPTER 14

experience,] but I'll say something in class and my mentor teacher will giggle, but then I'm like...the other day I tried to use sarcasm with a student, and then I thought, 'great, I can already smell the emails coming in.' That's why I've gone out of my way to avoid getting an Instagram and stuff, because I can just see it being dangerous for my job hunt.

"It's Like Writing History on Twitter"

Even though her response was negative and Kate didn't plan to continue using social media to construct a professional identity, she did see benefit in its capabilities, such as hashtags over time and across classes; she just chose not to use them:

In our class, I think we used it because it was maybe a way to connect to students on a different platform. It's not like, 'let's Canvas message.' And the messaging stuff is kind of easy to use on there. But also no matter how much I don't like to use Twitter, just because it annoys me, it's cool that you can click on a hashtag and just see all the things in there. And I think that's a good way to connect even across classes. You [as the instructor] have a running—it's like writing history on Twitter. You have years and years and years of these classes, and you said, you can go see what they did. I see why you used it, but I am not the person to go and see what they did.

Forming an Identity: "This Is Going to Be the Person That I Want to Portray Myself As"

Identity was a concept that Kate felt took time, and occurred in response to her interactions with former teachers, parents, and friends. She hinted at former challenges that shaped who she had become and that had made her want to be seen as positive and happy, and help others feel that way as well:

You don't just form it [identities]...I suppose you could, but I feel like identities are created, they're not just formed. I didn't wake up one day and decide I'm going to be the witty loud girl, it came out of years of people like former teachers yelling at me and being like, 'you're so quiet!' And so I thought, 'okay, screw you, I'm going to be loud!' It's a building-up process to define your true identity. It comes out of hardships and lows and highs and you decide, 'ok, this is going to be the person that I want to portray myself as.'

[I wanted to have this identity because] Pops was always kind of funny, but also a [jerk]...I just felt like I've seen some things that I don't want...if I have children I don't want them to do that kind of stuff or have any of that stuff happen. So I just decided that all of that, the cliché of I'm going to be better for it. As a person I'd rather be happy and make people around me happy than be down and dismal and upset all the time. That just doesn't make me feel good. I like making people laugh. My dad did it the same kind of way…I just watched him go through a bunch of crap, but he would always have a joke to tell. So I saw a lot of it and decided that's what I was going to be.

Kate noted some of the possibilities for creating an identity on social media; she focused on personal (cat videos) and not professional personae that one might encounter.

> If you're getting intense into using social media, you could try to figure out someone is you could look at their likes, and if they're in a relationship or not, or when they were born if you're into that 'I'm a Capricorn' thing. And just based on their posts, if they're steady posts and like cat photos and shout out to their mom, you can kind of figure out what kind of person they are and what kind of identity they're not afraid to show other people. If you're posting it to social media, you probably think that's something that it's okay for you to define yourself as. Not that a cat video defines you, but 'this has my name behind it.'

There were various interests that Kate acknowledged that she should not share on social media, which recalls her former concerns about job security and a general dislike of professional pursuits on social media. She monitored her online presence by removing tags from certain photos and avoiding certain topics:

> I like relatively perverted bathroom humor, but at no point am I going to post that. I've been tagged in some things from friends back home and I'm like, "that's a hide, no, you can't do that." Something I might normally laugh at, but it's not something I want my name behind.

> I'm not going to talk about anything sexual or anything on social media, I guess my sexual orientation I wouldn't be worried about, but like making crude jokes isn't something that I'm going to portray myself as. I posted a photo—see our flag [she points to a rainbow flag]—I said…it's never too late to say "love won," because that was when the White House and the gays were getting their rights and stuff. So that was one time I was like, 'boom!' and putting out my opinion.

EXTENSION ACTIVITIES

1. Identify some key aspects of persona that Kate wants to share on social media. What in her narrative suggests that she might value these things?
2. Develop interview questions to encourage a participant to tell a story about your research topic. Search for sample question formats online that can elicit lengthy, interesting responses.
3. Shadow a student in one of your classes and create a case study to augment your mini study data collection.

CHAPTER 15

CROSS-CASE ANALYSIS AND DISCUSSION

After a holistic review of the cases, various common themes emerged. These themes are outlined and discussed in this chapter and took the form of various interpretations of how teachers can and can't (or should and shouldn't) present themselves publicly. This chapter includes relevant survey findings to enhance the cross-case analysis.

While Twitter was the main focus of the participants' teacher education courses when it was required, other forms of social media such as Facebook, Instagram, and Yik Yak arose often in the participants' responses. All agreed that social media facilitates communication and connections with others, including scholars in the field, but views on whether those connections—professional or otherwise—were valuable or even interesting varied dramatically. The following sections describe common themes from the cross-case analysis and the findings of the survey. These common themes included social media as a way of life, tensions and opportunities, and identity formation.

A WAY OF LIFE

The data for this study showed that social media was simply a way of life for these beginning teachers. Whether they hated or feared aspects of social media, only wanted to use it personally, or embraced it completely for both personal and professional uses, every participant was familiar with social media and noted ways that it had impacted their lives both positively and negatively. Recalling the conceptual framework, the participants did use social media in either expressive or informational ways (Tufekci, 2008) and either lived as residents in online spaces or visited them occasionally (Wright et al., 2014).

Level of Prior Experience and Time Investment

Survey results suggested that time of use was a factor in the participants' enjoyment of Twitter for teacher education courses. Results were not statistically significant and the sample size for this context was small, but seven survey respondents who used Twitter for ten to 30 minutes per week enjoyed using Twitter, as compared to ten who did not enjoy using Twitter and who used it for less than ten minutes per week. Of course, there is no way to tell whether those who already disliked the idea of Twitter simply did not want to use it as often, or if those who invested more time in the activity liked it more to begin with, but this finding suggests that the idea

CHAPTER 15

of resident and visitor and expressive and informational uses from the conceptual framework (Wright et al., 2014) applied to these participants.

A striking difference between Nora and Callie or Kate was clear in the amount of time that they had chosen to invest in social media over the years. While it had been a way of life for Nora since high school, where she even had a secret Twitter account with one friend, Kate described using sites like Facebook only occasionally. Nora and Marina were residents on Twitter, and Callie and Kate were visitors. However, Callie's description of her interactions on Instagram suggested that she was a resident in that context.

Another aspect of the conceptual framework involved uses of social media from informational to expressive. Along with being residents, Nora and Marina also used Twitter in both expressive and informational ways; they formed identities through the posts that they constructed and used social media to gather information about topics of interest. Callie and Kate did use Twitter to gather and share information as it related to their courses when it was required, but it was not a memorable or important piece of their overall professional identity development. Once again, Callie used Instagram to express herself and her personal interests and did not consider Twitter to be a place where she had formed her professional identity.

Audience Awareness: Who Is Reading What I Am Writing?

Another common theme in the case narratives involved the ways in which the participants considered their audiences and those audience's expectations. Nora described the ways that she had honed her thinking about public posts, which began when actual members of her audience—her coaches, most often—responded to her choices by telling her to "watch that" because "you can't take it back." Nora had the intensive involvement of adults—her soccer coaches—as she posted publicly even before she became a teacher education student, but this was not the case for any other participants.

The sense of permanence and the public aspect of social media led to a real sense of fear and risk among all of the participants about presenting an unacceptable version of one's self, a self which could even lead to trouble when it came time to look for a job. Marina attributed this to former class discussions about the risks of using social media. Both Nora and Marina expressed the feeling that anything they posted on social media would be permanently linked to them. Furthermore, Kate acknowledged that she had already had trouble when interacting with earlier practicum teachers in person and that adding the element of policing herself on social media was simply more than she wanted to handle. Callie echoed Kate's sense of exhaustion with the sheer amount of work and time that accompanied keeping up with professional social media as well as school, friends, and other personal interests.

Additionally, as the participants used social media for new purposes and negotiated private, personal, and/or professional accounts, they feared that their usual audiences, which were generally made up of friends, would not be interested in their newly professional posts. Until the introduction of social media as a course requirement, the only audience that three of the participants had experienced was one composed of friends and family; Nora was the sole exception. Nora noted that she began using Twitter in a semi-professional manner before using it for a course, and also viewed her audience more broadly because it included members of her church and her various soccer communities. Indeed, the active nature of the communities in which Nora interacted seems to have sparked both a love of the format and an awareness of how she could present a professional persona online. Nora did not have the sense that she was shouting into a void where no one would hear her, although she did long for earlier years when she used social media as an outlet and could post more personal thoughts.

TENSIONS AND OPPORTUNITIES

Tension One: What Is Social Media's Purpose?

A tension emerged in the narratives: what the purpose of social media was, and whether it was a pleasant, non-academic distraction, as Marina noted, or whether it was a method of connecting with scholars and developing one's professional identity. The latter purpose had an accompanying feeling of falseness for some participants, and some (especially Kate and Callie) rejected what they saw as the idea that they had to present a fake public self. Perhaps ironically, though, Kate also noted that she only wanted to post updates that were positive, even though she acknowledged having been shaped by early negative experiences. She simply chose not to dwell on the negative because she didn't want others to view that as a part of her identity. Kate was also concerned with her effects on her audiences, and wanted them to feel positive as well. For Callie and Kate, who did not use Twitter as residents or in expressive ways, it was due to a conflict between what they saw as the purpose of social media and the tasks that they were asked to do for the course.

Balancing Capabilities, Expectations, and Frustration

Twitter and other forms of social media have different capabilities; generally, the case study participants noted that Facebook provided a space of connection for their family members, including their parents. The character limit and focus on Twitter were points of frustration for Callie, as well as for 28% of the survey respondents. Callie also preferred sharing images and named Instagram as a preferred site for doing so; indeed, Instagram features image editing tools and lenses that are not present on Twitter or Facebook.

All of the case study participants noted that Facebook was generally used among older people such as their parents or extended family members. Callie had been invited to and joined an art teachers' group on Facebook by a mentor teacher. Instagram was a frequently used site for two participants (Callie and Marina), who noted that it was better for image sharing, but not for sharing with larger audiences. Kate observed that anyone who wanted to could share their views on social media, even if those views were narrow-minded or offensive, and this was an aspect of social media that made her less interested and unlikely to participate. Even though she considered herself skilled at presenting a professional identity on social media, Nora noted that she sometimes wished that she could let her guard down and write in more honest ways. The idea that more time spent viewing others' posts or highlight reels could lead to jealousy and depression echoed the findings of Tandoc et al. (2015).

Another capability of Twitter was to connect to scholars and broaden student users' audiences, and some participants did acknowledge that they should be aware of the possibilities of social media for teaching and learning. Nora noted that it was a common practice in her program to experience something that she might use later in her own teaching. Callie was inspired and the believed that teachers should "start a movement to mirror how students are learning outside of school." Given Callie's preference for Instagram and her content area of art, the findings of Milton's (2015) study of Instagram in a secondary social studies class suggest that she may find success with similar projects in the art classroom. Milton (2015) found that

> Students (1) felt that social media is valuable in an educational setting, (2) felt that combining their use of social media with specific geography content was meaningful, (3) were able to make connections between their local communities and social studies content, and (4) found that sharing their interpretations of social studies content through images was a meaningful way to learn. (p. 40)

Tension Two: Becoming a Teacher

The participants in this study experienced the various tensions associated with becoming a teacher. They lived in two worlds: that of the college student and that of the teacher, and developed personae for each setting; these settings included both face-to-face and online spaces. Kate described herself as silly and "foofy" in her classes, but did not feel that that she could not be this way during her practicum placements. Studies show that young adults consider aspects such as whether they present "real" or "false" selves on social networking (Michikyan, Dennis, & Subrahmanyam, 2014), and this was salient for these participants as well. Findings in Davies (2012) were also relevant to this tension: the chance to post details about what one is doing when not in the presence of friends allows individuals to seem more authentic and demonstrate that they are certain kinds of people. The requirement to post about course-related topics outside of class time, but in a personal style—which is a standard way of interacting on social media—was jarring, fake, or unnatural for some. Kate, in particular, resisted the idea of a fake self, which she equated with a more professional presence with which she did not identify. In other words, it is one thing to contribute to class conversations or be seen as a teacher in a practicum or classroom setting; it is another to enact the role of the teacher in online spaces, where one may not feel like enacting that role at all times.

Recent research has investigated the affordances of Twitter and found that students engage in new literacy practices online (Davies, 2012; Greenhow & Gleason, 2014). Additionally, many teachers who use Twitter tend to use it for personal purposes and then gradually begin to use it for professional purposes such as sharing resources, joining educational chats, and building a PLN, especially with those who may not otherwise be accessible (Krutka & Carpenter, 2014). This process of moving from the personal to the professional was indeed the case for Nora and Marina. However, the blend of personal content and professional content varied a great deal: Nora penned thousands of tweets, many of them about personal events in her life, while Marina and, to a lesser degree, Kate, and Callie, decided to allow the professional to remain the only kind of content on her public Twitter profile. Kate and Callie also maintained other sites such as Facebook or Instagram more actively, where they chose to post more personal content and interact with friends and family. In contrast, Nora described Facebook as more outdated and as a place where she did not choose to invest her time.

It is likely the case that some participants felt frustrated by the need to fracture their social media efforts to yet another site or another purpose. The participants were encouraged to use Twitter professionally, in some cases when they did not wish to use it (or even any social media) at all.

PERSONA AND IDENTITY FORMATION

It should come as no surprise that many people, including several of the participants in this study, enjoy and interact on social media in various ways; the issue of identity development in online spaces is less clear. Returning to the conceptual framework, sources for persona and identity formation include personal context (upbringing, prior experiences, and teacher training) and classroom and school contexts (course goals, instructional levels and social interactions). These components figure into the ways that preservice teachers used signals to establish personae, which then led to emerging identities. This framework was developed and refined with a similar population of preservice teachers.

Contexts

Our experiences inform how we develop and present our various selves, and this was true for the participants. Kate alluded to challenges from her childhood as formative experiences that she would not want others (especially children) to have to experience, and that informed her choice to present only positive or humorous messages on social media. Similarly, her choices about using social media to present a professional identity were also related to her prior negative experiences with

CHAPTER 15

practicum settings and mentors. She felt that if she were to enact her authentic self on social media, that she might get into trouble or even lose a job in the future, so she did not use it often and viewed it as something more suited to celebrities.

At the other end of the spectrum, Nora had experiences in her youth where coaches would review her performance both on and off the field and who would recommend certain kinds of posts over others on social media. Nora wanted to form a professional identity for teaching as well as for church and soccer.

Marina noted that her parents had steered her away from certain peers by switching her public school when she was a child. Marina was aware of audiences for her interactions, and this affected what she chose to post on social media when she described deleting images with any alcohol at all, even though she was of the legal drinking age and she had been at a tasting event with her family.

Callie, who grew up loving art, enjoyed sharing images, so she preferred the format of Instagram to Twitter or Facebook.

Although research has found teacher models to be important to how novice teachers construct identities (Flores & Day, 2005), the actions of the instructor to model how to form a professional identity while interacting on social media were not a source that the participants named. Furthermore, interactions with K-12 students can affect teachers' presentations of self (Flores & Day, 2005), but none of the participants interacted with students on social media due to warnings and concern about maintaining professional boundaries.

Views about How and Why Identities Are Constructed

Identity is a complex concept and there is no single definition on which all scholars agree. The survey and case study participants had considered how identities were formed ideas and believed that the process involved their actions, interactions, and experiences. Furthermore, the majority of survey respondents agreed that it is possible to construct an identity on social media through one's posts and interactions. One survey respondent even likened identity creation on social media to "curating" what others view. Indeed, the metaphor of a museum is apt and reveals the care with which some may craft identities in online spaces; it also shows why several survey respondents and case study participants felt that constructing identities online can seem false. Case study and survey participants noted concerns about the highlight reel, where they may be comparing their own lives to an incomplete picture of someone else's, which can be depressing or stressful. Callie identified a complex view of how and why people think about presenting selves online when she stated:

> Sometimes I just want to tell people to cut the shenanigans and show us your real life, but I don't, because that would be the image I then made of myself. It's a lot different to tell someone that in person; it's not recorded as a permanent document as it is on social media. What we say on social media can always be back-tracked and used against [us] for whatever reason.

This statement showed a complication of social media; even though it provides a space for asynchronous interactions and "skilled reading of the situation" (Davies, 2012, p. 27), through comments and "likes," social media also has a permanence and publicity that lead to considerations about whether it is worth writing messages for all to see.

Only two of the participants, Nora and Marina, described feeling as if they were deliberately and intentionally building a professional identity on social media. Some, like Kate, even resented or at least rejected being asked to use social media in these ways. Callie indicated that the falseness of social media could be depressing and stressful. However, the participants had all developed strategies to control the identities that they formed on social media, such as deleting tags and comments that could be misconstrued. Furthermore, persona and identity development were not goals that these preservice teachers set out to achieve (indeed, they were not actually asked to do it; it was an intended by-product of the use of Twitter), whether on social media or in other settings; rather, the opportunity and need to construct professional identities arose when sparked by an outside source. Developing one's selves is generally more casual or even accidental; it is the accumulation of these moments and the formation of our audiences—whether intentional or not—that build each of us into the teachers we become.

Professional Identity Creation: Is It Asking Too Much?

At the heart of this study, including the conceptual framework on which it was developed, is the idea that identity can be constructed on social media. While Nora and Marina did note that they were making purposeful choices to be seen as professional, the survey suggested that nature of the preservice teachers' thinking did not extend to explicit constructions of professional identity: many simply interacted online. In most cases, they interacted online for the course because it was required. Constructing a professional identity was a step beyond what was assigned for the course (sample Twitter assignments included use the hashtag, share resources, tag a scholar, etc.), so it is unsurprising that not all participants thought of Twitter as a place where they could or even wanted to construct a professional identity. Furthermore, while there was not explicit in-class instruction in how to construct identities online, it certainly could be the case that such instruction may not have been effective because transformation cannot be imposed (Veletsianos, 2011).

A common culminating assignment in teacher education programs, including the one that these preservice teachers attended, asks preservice teachers to complete a portfolio. The portfolio is often comprised of assignments and written reflections from the students' time in the program. Portfolios generally are not viewed as identity creation spaces. Twitter can be such a space, but was not for the survey respondents and all of the case study participants, with the exception of Nora and, to a lesser degree, Marina. In practice, purposeful professional identity creation on Twitter only occurred

CHAPTER 15

occasionally, perhaps because of the habits that these young adults have developed around online interactions or perhaps because the participants did not wish to identify publicly as teachers. It may be the case that one's Twitter account could dovetail with the portfolio by adding a more public space for original lessons, units, and writing such as blog posts or reflection papers; in this way, the work of the program and of connecting with others in the field could be more aligned and purposeful.

Regardless of the methods adopted that might increase the opportunities for preservice teachers to share their work and thinking more publicly, it is still the case that forming an identity, professional or otherwise, is complicated. The preservice teachers in this study existed in two worlds: the public school and the university. They had differing ideas about identity and how, why, and whether they might share it on social media. While the intent of the assignment to use Twitter was not to have a certain kind of identity, these preservice teachers were well aware that their public identities mattered. They wanted to be hired as full-time teachers in the near future. An awareness of their particular interests, needs, and fears regarding social media was a consideration when the instructor interacted with the participants in courses; it is likely that even more time must be devoted to these areas for the future.

SUMMARY

- Use of social media was a way of life for participants, especially for their personal lives.
- Case participants were evenly divided along both kinds of interactions (expressive to informational) and time of use (resident to visitor).
- Awareness of audiences was important to how the participants thought about their posts and interactions online.
- Some tensions for the participants included questions about the purpose of social media, feelings of falseness or frustration at what was possible on Twitter, and concerns about becoming a teacher and interacting in public spaces.
- Contexts and experiences from childhood were integral to how the participants interacted on Twitter.
- In practice, professional identity creation occurred occasionally for two of the four case study participants, even though it was not explicitly taught.

EXTENSION ACTIVITIES

1. Review the cases and survey findings from the previous chapters. Do you notice any additional themes not mentioned here?
2. Identify the biggest challenges and opportunities for you when it comes to social media in education. Have the findings of this study expanded or changed your views in any way?
3. Identify themes across the data you collected for your own study. How does what you found compare to what you expected to find? What is surprising?

CHAPTER 16

IMPLICATIONS

The survey, tweets, interviews, and case studies from this study have provided a window into the ways that preservice teachers think about and use social media, especially as it relates to their personae and future professional identities. The study had various implications for beginning teachers, teacher educators, and future educators; these are outlined in this chapter.

IMPLICATIONS FOR BEGINNING AND PRACTICING TEACHERS

Beginning Teachers

Living at the intersection of student and teacher, no matter the setting, can be complicated and stressful. When it comes to navigating online interactions and identities, there are several recommendations that follow from this study's findings. Due to the challenges associated with maintaining personal and professional relationships in a public space, preservice teachers may have more success in both areas if, rather than keeping separate Twitter accounts (one personal, one professional), they use certain sites for personal connections, and Twitter for professional connections. This is due to the confusion and complications that resulted when preservice teachers in this study attempted to use two separate Twitter accounts. As they have likely been warned, preservice teachers should remain vigilant about their posts and avoid offensive content or interactions. However, new teachers should not confuse vigilance with avoidance. Knowing about and using social media for education is a way of life for people of all ages, and teachers must stay informed and knowledgeable about current developments. While the participants for this study were preservice teachers, which is appropriate because 90% of young adults use social media (Pew Research Center, 2013), there are aspects of the findings and from the literature that can inform the work of practicing teachers as well.

Practicing Teachers

Professional development is a mixed bag for teachers; sometimes it is useful and reignites one's passion for teaching. When poorly planned or implemented, teachers may view professional development activities as a waste of time or even detrimental to their teaching. The benefit of using Twitter for professional development is that it can be done in short bursts of time. Location is less of a limiting factor as compared to face-to-face professional development courses; all manner of scholars can be accessed on Twitter. However, care in evaluating and verifying online sources must be present when using social media for professional development. Information can always be outdated, incomplete, or incorrect, and followers may not always be who they seem.

The value of social media extends beyond connecting with peers and colleagues, and even beyond its strength as a source of self-directed professional development: teachers can use social media to construct a digital identity that will model ways of interacting in the world for their students. From sharing conference pictures to conversing with scholars and posting informational videos and links, teachers who use social media in these ways become models for the next generation of tech-savvy students.

Finally, the findings indicate that there is a series of questions that teachers of all kinds—preservice, new, and practicing—must ask themselves when they consider how they present public selves; these include the following:

- In what settings am I a teacher?
- What does it mean to enact the role or persona of the teacher?
- When am I someone else?
- From where did I get my ideas about how to present myself as a teacher and in other roles?
- How similar and different are my various selves?
- Why do I blend or separate these selves, and should I continue to do so?

Attention to the above questions in written and spoken reflection will contribute to teachers' comprehensive professional and personal identity development. Blogs may be an especially useful format for those teachers who are comfortable making their thoughts public in an online space.

IMPLICATIONS FOR TEACHER EDUCATORS

Implications of the study in this text for teacher educators include increased use of Twitter to connect students within and across programs to encourage peer modeling of professional identity development, and inclusion of related work such as blogging (which can be shared with a wider audience on Twitter) to encourage preservice teacher development. The survey and case study data for this study show that rich and robust class discussion can result if topics include thoughts about methods of identity formation in everyday life, including on social media.

Because social media is a way of life that we cannot avoid or ignore, it is crucial that teacher educators adjust their methods to take advantage of the format's possibilities. The experiences of the participants support the use of Twitter for communication, provided there is some attention to explicit modeling and training of how to use the sites for maximum effect. Time is another factor to consider when including social media as a course requirement. A goal of 10 to 30 minutes of use per week may elicit more positive responses from preservice teachers than shorter amounts of time. Again, recommendations include that modeling of useful strategies such choosing followers, reviewing the feed, using hashtags correctly, and retweeting strategically will support new teachers' enjoyment and success with the format.

While other social media sites such as Instagram and Facebook have strengths and should not be overlooked, in this study the value that came from connecting on one site would have been impossible if users had been divided among their preferred sites. Twitter encouraged users who were not anonymous and who could share personal details, resources, and communications in a user-friendly format. Concerns about privacy continue for many preservice teachers, so methods of masking one's actual identity may be shared to alleviate some of these concerns, such as using a first and middle name for a social media account instead of a first and last name. Course hashtags to consolidate responses helped build community and enabled users to share resources and reactions more publicly and access former students' responses. In order to demonstrate the possibilities of the format, teacher educators, university supervisors, and mentor teachers may share examples and non-examples of different kinds of posts with the purpose of analyzing their effects on audiences.

IMPLICATIONS FOR RESEARCHERS

This study provided a qualitative view of one particular context and four preservice teachers, and adds to the growing body of research on social media in education and for identity development. Future studies should include direct observation of participants as they use Twitter and other forms of social media, both in the short and long term. Considering both how, why, and when users interact online and what they think about those interactions will offer a more complete picture of how teachers learn and live online.

Social media does not simply mean checking Facebook or Twitter on one's computer; these sites are accessible from any smart phone as well. It was beyond the scope of this study, but it is likely that whether one accesses social media on a mobile phone or on a computer affects the kinds or amount of comments, images, and interactions that result. It is also possible that the participants responded negatively to use of Twitter, but would (or already do) enjoy using other sites more for professional identity development.

Viewing the actual tweets of the case study participants in this study supplemented their interviews and provided a more rounded case portrait; further research into actual teachers' posts and interactions will continue to deepen our understanding of the ways that beginning teachers learn and live online. The strategy of the "walk through" of a participants' social media site as used in Davies (2012) may be useful for future studies of beginning teachers' actions and interactions on social media. Wright et al. (2013) mapped the networks of students who interacted on social media (in their case, Google+). Similar visualizations of such data in future studies can continue to make the interactions between and among teachers and students more evident to outside audiences. Finally, scholars who investigate Twitter and other

CHAPTER 16

forms of social media must continue to determine the effects of publicly shared messages on their audiences. When we lurk, like, or listen to the constant stream of online conversation, it affects us in ways that have not yet been fully documented in research.

SUMMARY

- Teachers should take care and be aware of their audiences when presenting themselves online, but also take advantage the many opportunities for professional development available on Twitter and other social media.
- Teacher educators should make identity formation a regular topic of discussion and reflection.
- Teachers need models of how and why to use social media to develop a professional identity.
- Researchers should observe the ways that participants use and think about social media, including how and why they interact online.

EXTENSION ACTIVITIES

1. Consider your study's design, including the data and population. Identify any limiting factors. What would you change if you had more time or choice about data formats, population, setting, and other aspects?
2. Respond: given the results of your study, what should teachers know and do? What other topics should researchers investigate now that you have answers to your initial research questions?
3. After reviewing the data for this study of preservice teacher identity development on Twitter, what do you believe are some additional recommendations that might be made for beginning teachers, teacher educators, or researchers? Are there other groups that may also be affected by these findings?

CHAPTER 17

CONCLUSIONS AND NEXT STEPS

> There is no longer a public self, even a rhetorical one. There are only lots of people protecting their privacy, while watching themselves, and one another, refracted, endlessly, through a prism of absurd design.
>
> (Lepore, 2013, para. 27)

Because Twitter and other social media sites are relatively new at this point in time, their applications for education and their roles in our lives are still developing. As noted in the previous chapter, research into how students, teachers, and scholars use Twitter and other forms of social networking is of paramount importance as more teachers turn to social media outlets for professional development, including persona and identity development—even if those public selves are not developed intentionally. Audiences are all around us, in every setting. When we comment on a friend's post on Facebook, people we have never met—our friend's friends—are reading our words; sometimes they even interact with us. At this time in history, when young adults already consider whether they present "real" or "false" selves on social networking (Michikyan et al., 2014), teachers must be aware of how they are guiding students of all ages to present certain kinds of selves. Teacher or peer feedback on presentations of self may have once occurred in the private spaces of student writing in the classroom, as when I viewed a mentor teacher's incredulous comment on a student's inclusion of curtain color choice in an essay about life goals. Now, writing and feedback has become much more public, and, for better or worse, students can get feedback from all manner of audiences. Just as users must beware of the highlight reel, they should also beware of the echo chamber.

Concerns abound about how and whether teachers should utilize social media. Lessened privacy and a potential minefield of miscommunication or missteps are all potential issues for anyone who uses social media, including new teachers. There is founded concern about job security. A plethora of challenges—from a need for technological knowledge, to simple access to the internet or blocked sites, to knowledge of what sources of professional development are most useful or accurate—exist in the current digital landscape. It would be easy to warn new and preservice teachers away from using social media. It is easy to tell teachers that the risk is not worth it. But that would be wrong. Similar to most other controversial topics in education throughout history, the stance that teachers must take is to maintain an open mind and a willingness to learn. As Jim Burke notes in *The English Teacher's Companion*,

CHAPTER 17

> Let me say first that nothing can teach you more about what digital writing is and how to do it than writing such texts. We do not have to *like* Twitter or Facebook, but we have a professional obligation to understand and keep in mind how written communication is evolving in the world for which we are preparing our students. (p. 129)

Framing awareness of social media as a professional obligation highlights just how serious it is; teachers must attend to the ways that students communicate in order to develop and deliver appropriate instruction to suit their needs and engage their interests.

Allow me to return for a moment to this text's secondary purpose, which appeared throughout the chapters in the form of extension activities designed to walk beginning teachers through the process of researching topics of interest with the goal of improving their teaching practice. This book has served this purpose for me. To simplify the process of inquiry, here are the steps I took to determine if an idea was working in my classroom. In roughly the following order, I:

- learned about an idea that I wanted to try in my classroom, so I built it into my course
- developed a question to determine the impact of the new method
- created a plan to collect data so that I could know more about students' thoughts
- analyzed the data to find common themes
- wrote this text to summarize the entire process
- reflected on my teaching as a result of the process
- adjusted my plan for the next year based on my findings, the process, and my reflections

The process was lengthy, but it vastly improved my practice. I can now provide examples of actual student responses in my future courses if I opt to implement this method again. Furthermore, even more than answering this one question, my goals are to demonstrate how this process works for new teachers so that they can enact similar methods in their own courses. Will they publish books or articles? Possibly a small fraction of them will, and generations of new teachers will learn from those texts as well. What was the alternative to action research, when teachers implemented new teaching methods or practices in the classroom? They might have had a feeling that students liked it, or they might have shared the idea with teachers in their school or district—or maybe a wider audience at a conference. The benefit of action research is that it provides qualitative and quantitative data to demonstrate teachers' impact on students that go far above, beyond, and deeper than scores on high-stakes, standardized tests.

At a time when students—both at the K-12 and college levels—may use forms of social media without careful consideration and without the knowledge of their parents or teachers, it is essential that teacher educators be aware of how these tools are used in order to determine best practices and support the identity formation that can and should occur in this public space.

EXTENSION ACTIVITIES

1. Construct a plan to develop your own identities—both personal and professional or a blend of the two in one space—on social media.
2. Plan a lesson or unit for K-12 students or adults to convey what you believe are the best practices of how and why one should interact with others online.
3. Reflect on the process of developing your study, collecting and analyzing the data, and sharing the results. What would you do differently, if you could start again at the beginning?

REFERENCES

Aarts, O., van Maanen, P. P., Ouboter, T., & Schraagen, J. M. (2012, December). Online social behavior in Twitter: A literature review. In *Data Mining Workshops (ICDMW), 2012 IEEE 12th International Conference*, 739–746. doi:10.1109/ICDMW.2012.139

Anderson, G. (2015, October 13). Re: Tough teacher [Online forum comment]. Retrieved from https://ncte.connectedcommunity.org/communities/community-home/digestviewer/viewthread?MID=24294&GroupId=757&tab=digestviewer&UserKey=fd4e788e-64eb-4f21-bcb7-0161ca924a39&sKey=3E1F1A354E134551B9FD&auth=b87g7140dc80a540g5216g94hg1204a8979-0bf5448651d93d88#bm7

Arendt, H. (1981). *The life of the mind.* Orlando, FL: Harcourt, Inc.

Bhabha, H. K. (1994). *The location of culture.* London, England: Routledge.

Birnbaum, M. (2008). *Taking Goffman on a tour of Facebook: College students and the presentation of self in a mediated digital environment* (Doctoral dissertation). Retrieved from http://arizona.openrepository.com

Bollen, J., Mao, H., & Pepe, A. (2011). Modeling public mood and emotion: Twitter sentiment and socio-economic phenomena. *Proceedings of the Association for the Advancement of Artificial Intelligence, 11*, 450–453. Retrieved from http://www.aaai.org/ocs/index.php/ICWSM/ICWSM11/paper/viewFile/2826/3237/

boyd, d. (2014). *It's complicated: The social lives of networked teens.* New Haven, CT: Yale University Press.

Brown, T. (2006). Negotiating psychological disturbance in preservice teacher education. *Teaching and Teacher Education, 22*, 675–689. doi:10.1109/ICDMW.2012.139

Bruner, J. (1991). The narrative construction of reality. *Critical Inquiry, 18*(1), 1–21. Retrieved from http://www.jstor.org/stable/1343711

Carpenter, J. P., & Krutka, D. G. (2014). How and why educators use Twitter: A survey of the field. *Journal of Research on Technology in Education, 46*(4), 414–434. doi:10.1080/15391523.2014.925701

Cavanagh, M., & Prescott, A. (2007). Professional experience in learning to teach secondary mathematics: Incorporating pre-service teachers into a community of practice. *Mathematics: Essential Research, Essential Practice, 1*, 182–191.

Coby, J. (2013, April 29). *Sixteen sassy tweets from the nation's 16th largest school district.* Retrieved from https://www.buzzfeed.com/jonathancoby/16-sassy-tweets-from-the-nations-16th-largest-sch-afb2

Common White Girl. [girlposts]. (2016, June 22). aw thank you [Tweet]. Retrieved from https://twitter.com/girlposts/status/745708871616962560

Cuddy, A. (2012, June). Amy Cuddy: Your body language shapes who you are [Video file]. Retrieved from https://www.ted.com/talks/amy_cuddy_your_body_language_shapes_who_you_are?language=en

Cuenca, A., Schmeichel, M., Butler, B., Dinkelman, T., & Nichols, J. (2011). Creating a "third space" in student teaching: Implications for the university supervisor's status as outsider. *Teaching and Teacher Education, 27*, 1068–1077. doi:10.1016/j.tate.2011.05.003

Dabbagh, N., & Kitsantas, A. (2011). Personal learning environments, social media, and self-regulated learning: A natural formula for connecting formal and informal learning. *Internet and Higher Education, 15*(1), 3–8. doi:10.1016/j.iheduc.2011.06.002

Davies, J. (2012). Facework on Facebook as a new literacy practice. *Computers & Education, 59*(1), 19–29. doi:10.1016/j.compedu.2011.11.007

Davis, J. S. (2010). *Who am I this time? Secondary preservice teachers: Developing and presenting a teaching persona* (unpublished doctoral dissertation). University of Virginia, Charlottesville, VA.

Davis, J. S. (2013, February 4). The super bowl, online persona, and…you? [Web blog post]. Retrieved from http://blog.janine-davis.com/uncategorized/the-super-bowl-online-persona-and-you/

Davis, J. S. [JanineSDavis]. (2013, September 25). #535Davis and #351Davis--Tweetdeck (available through Chrome) is free and great way to organize your Twitter hashtags [Tweet]. Retrieved from https://twitter.com/JanineSDavis/status/382859019343261696

REFERENCES

Davis, J. S. (2014, April 15). Social media as third space for teacher education [Web log post]. Retrieved from http://blog.janine-davis.com/uncategorized/social-media-as-third-space-for-teacher-education/

Davis, J. S. [JanineSDavis]. (2016, June 7). http://getschooled.blog.myajc.com/2016/05/27/university-of-georgia-professor-explains-his-aspergers-advantage-and-disabling-assumption-of-disorder/ ... #TOEDavis from our text's author! #351Davis and also thought of you and your classes, @Techtweed and @Jen_D_Walker [Tweet]. Retrieved from https://twitter.com/JanineSDavis/status/740153675914547200

Davis, J. S. [JanineSDavis]. (2016, June 15). #545Davis #TOEDavis #writinggoals [Tweet]. Retrieved from https://twitter.com/JanineSDavis/status/743069677820256257

Derks, D., Bos, A. E., & Von Grumbkow, J. (2007). Emoticons and social interaction on the Internet: the importance of social context. *Computers in Human Behavior, 23*(1), 842–849. doi:10.1016/j.chb.2004.11.013

Edutopia. (2015). *Social media: Making connections through Twitter*. Retrieved from http://www.edutopia.org/practice/social-media-making-connections-through-twitter

Erikson, E. H. (1968). *Identity: Youth and crisis*. New York, NY: Norton.

Ferguson, K. (2012, June). Kirby Ferguson: Embrace the remix [Video file]. Retrieved from https://www.ted.com/talks/kirby_ferguson_embrace_the_remix?language=en

Flores, M. A., & Day, C. (2005). Contexts which shape and reshape new teachers' identities: A multi-perspective study. *Teaching and Teacher Education, 22*(2), 219–232. doi:10.1016/j.tate.2005.09.002

Gee, J. (2000). Identity as an analytic lens for research in education. *Review of Research in Education, 25*, 99–125.

Gleason, B. (2016). New literacy practices of teenage Twitter users. *Learning, Media and Technology, 41*(1), 31–54. Retrieved from http://www.tandfonline.com/

Greenhow, C., & Gleason, B. (2012, October). Twitteracy: Tweeting as a new literacy practice. *The Educational Forum, 76*(4), 464–478. doi:10.1080/00131725.2012.709032

Gunraj, D. N., Drumm-Hewitt, A. M., Dashow, E. M., Upadhyay, S. S., & Klin, C. M. (2016). Texting insincerely: The role of the period in text messaging. *Computers in Human Behavior, 55*(B), 1067–1075. doi:10.1016/j.chb.2015.11.003.

J. Paul Getty Museum [GettyMuseum]. (2016, June 22). To celebrate #musEmoji day, we're sharing our fav labors of Hercules using ONLY emojis. Guess which labor each emoji-story tells—good luck![Tweet]. Retrieved from https://twitter.com/GettyMuseum/status/745633138383142912

Kagan, D. M. (1992). Professional growth among preservice and beginning teachers. *Review of Educational Research, 62*(2), 129–169. doi:10.3102/00346543062002129

Karchmer-Klein, R. (2013). Best practices in using technology to support writing. In S. Graham, C. A. MacArthur, & J. Fitzgerald (Eds.), *Best practices in writing instruction* (pp. 309–333). New York, NY: The Guilford Press.

Knowles, J. G., & Holt-Reynolds, D. (1991). Shaping pedagogies through personal histories in preservice teacher education. *Teachers College Record, 93*(1), 87–113.

Kvale, S. (1996). *Interviews: An introduction to qualitative research interviewing*. Thousand Oaks, CA: Sage.

Lepore, J. (2013, June 24). The prism: Privacy in an age of publicity. *The New Yorker*. Retrieved from http://www.newyorker.com/magazine/2013/06/24/the-prism

Lortie, D. (1975). *Schoolteacher: A sociological study*. Chicago, IL: University of Chicago Press.

Major, C. H. (2015). *Teaching online: A guide to theory, research, and practice*. Baltimore, MD: JHU Press.

Manca, S., & Ranieri, M. (2013). Is it a tool suitable for learning? A critical review of the literature on Facebook as a technology enhanced learning environment. *Journal of Computer Assisted Learning, 29*(6), 487–504. doi:10.1111/jcal.12007

McBride, M. (2008). Classroom 2.0: Avoiding the "creepy treehouse" [Web log post]. Retrieved from http://melaniemcbride.net/2008/04/26/creepy-treehouse-v-digital-literacies/

Merilli, A. (2015, September 9). Socality barbie hits uncomfortably close to home. *The Atlantic*. Retrieved from http://www.theatlantic.com/entertainment/archive/2015/09/hipster-socality-barbie-shows-the-cliche-of-instagram-authenticity/404431/

REFERENCES

Michikyan, M., Dennis, J., & Subrahmanyam, K. (2014). Can you guess who I am? Real, ideal, and false self-presentation on Facebook among emerging adults. *Emerging Adulthood, 3*(1), 55–64 doi:10.1177/2167696814532442

Milton, J. (2015). *Picture a different world: Instagram's effect on students' attitudes toward social studies* (Unpublished master's thesis). University of Mary Washington, Fredericksburg, VA.

Murtha, J. (2015, September 15). How audience engagement editors are guiding online discussions. *Columbia Journalism Review*. Retrieved from http://www.cjr.org/analysis/before_many_americans_had_awoken.php

National Council for the Social Studies. (2013). *Technology position statement and guidelines*. Retrieved from http://www.socialstudies.org/positions/technology

National Council for Teachers of English. (2013). *The NCTE definition of 21st century literacies*. Retrieved from http://www.ncte.org/positions/statements/21stcentdefinition?roi=echo4-33272070595-80981468-952ff5a7b95647a2a7ba0934fe251d2e&utm_source=Digital+Literacy+2016-02-17&utm_medium=Email&utm_campaign=Books

Naveed, N., Gottron, T., Kunegis, J., & Alhadi, A. C. (2011, June). Bad news travel fast: A content-based analysis of interestingness on Twitter. *Proceedings of the 3rd International Web Science Conference, USA, 8*. doi:10.1145/2527031.2527052

Pew Research Center. (2013, December 27). *Social networking fact sheet*. Retrieved from http://www.pewinternet.org/fact-sheets/social-networking-fact-sheet/

Popova, M. (2015, October 14). *Hannah Arendt on being vs. appearing and our impulse for self-display*. Retrieved from https://www.brainpickings.org/2015/10/14/hannah-arendt-life-of-the-mind-being-appearing/

Popova, M. (2016). *About*. Retrieved from https://www.brainpickings.org/about/

Popova, M. (2016, January 6). *Beloved poet and philosopher Kahlil Gibran on the seeming self vs. the authentic self and the liberating madness of casting our masks aside*. Retrieved from https://www.brainpickings.org/2016/01/06/kahlil-gibran-madman-masks/?utm_content=buffer53cf6&utm_medium=social&utm_source=twitter.com&utm_campaign=buffer

Popova, M. (2016, March 2). *What makes a person: The seven layers of identity in literature and life*. Retrieved from https://www.brainpickings.org/2016/03/02/amelie-rorty-the-identities-of-persons/

Prensky, M. (2001). Digital natives, digital immigrants (Part 1). *On the Horizon, 9*(5), 1–6.

Rankin, M. (2009). *Some general comments on the 'Twitter Experiment'*. Retrieved from http://www.utdallas.edu/~mar046000/usweb/twitterconclusions.htm

Reeve. E. (2016, February 17). The secret lives of Tumblr teens. *New Republic*. Retrieved from https://newrepublic.com/article/129002/secret-lives-tumblrteens?utm_source=nextdraft&utm_medium=email

Saunders, S. (2008). The role of social networking sites in teacher education programs: A qualitative exploration. In K. McFerrin, R. Weber, R. Carlsen, & D. Willis (Eds.), *Proceedings of Society for Information Technology and Teacher Education International Conference* (pp. 2223–2228). Chesapeake, VA: AACE. Retrieved from http://www.editlib.org/p/27538

Seaman, J., & Tinti-Kane, H. (2013). *Social media for teaching and learning*. Boston, MA: Pearson Learning Systems.

Smithsonian Air and Space Museum [airandspace]. (2013, February 3). In the dark? Imagine being on Apollo 12 when they lost electrical power during launch: http://s.si.edu/hnNxh #MuseumSuperBowl [Tweet]. Retrieved from https://twitter.com/airandspace/status/298251903052169217

Suh, B., Hong, L., Pirolli, P., & Chi, E. H. (2010, August). Want to be retweeted? Large scale analytics on factors impacting retweet in twitter network. *Proceedings of the IEEE Conference on Social Computing, USA*, 177–184. doi:10.1109/SocialCom.2010.33

Tandoc, E. C., Ferrucci, P., & Duffy, M. (2015). Facebook use, envy, and depression among college students: Is facebooking depressing? *Computers in Human Behavior, 43*, 139–146. doi:10.1016/j.chb.2014.10.053

Tufekci, Z. (2008). Can you see me now? Audience and disclosure regulation in online social network sites. *Bulletin of Science, Technology & Society, 28*(1), 20–36. doi:10.1177/0270467607311484

REFERENCES

Unhelpful high school teacher. (n.d.), Retrieved from http://knowyourmeme.com/memes/unhelpful-high-school-teacher

University of Mary Washington. (n.d.), *A domain of one's own*. Retrieved from http://umw.domains/about/

Valkenburg, P. M., Peter, J., & Schouten, A. P. (2006). Friend networking sites and their relationship to adolescents' well-being and social self-esteem. *Cyber Psychology & Behavior, 9*(5), 584–590. doi:10.1089/cpb.2006.9.584

Veenman, S. (1984). Perceived problems of beginning teachers. *Review of Educational Research, 54*(2), 143–178. doi:10.3102/00346543054002143

Veletsianos, G. (2011). Designing opportunities for transformation with emerging technologies. *Educational Technology, 51*(2), 41. Retrieved from http://www.veletsianos.com/wp-content/uploads/2011/02/designing-opportunities-transformation-emerging-technologies.pdf

Veletsianos, G. (2012). Higher education scholars' participation and practices on Twitter. *Journal of Computer Assisted Learning, 28*(4), 336–349. doi:10.1111/j.1365-2729.2011.00449.x

Virta, A. (2002). Becoming a history teacher: Observations on the beliefs and growth of student teachers. *Teaching and Teacher Education, 18*(6), 687–698. Retrieved from http://www.leeds.ac.uk/educol/documents/00001556.htm

Wake County Schools. [WCPSS]. (2016, June 9). WE MADE IT! Please excuse us today if you see us dancing around the copy machine. #ohhappyday! [Tweet]. Retrieved from https://twitter.com/WCPSS/status/740863989672992768

Warlick, D. (2009). Grow your personal learning network: New technologies can keep you connected and help you manage information overload. *Learning & Leading with Technology, 36*(6), 12–16. Retrieved from http://files.eric.ed.gov/fulltext/EJ831435.pdf

Weber, S., & Mitchell, C. (1995). *That's funny, you don't look like a teacher: Interrogating images and identity in popular culture*. Philadelphia, PA: The Falmer Press.

Wells, L. (1994). Moving to the other side of the desk: Teachers' stories of self-fashioning. *American Council of Learned Societies Occasional Paper, (23)*. Retrieved from http://archives.acls.org/op/op23wells.htm

Whalen, Z. (2016, February 24). Notes on teaching with Slack [Web log post]. Retrieved from http://www.zachwhalen.net/posts/notes-on-teaching-with-slack/

Williams, J. (2014). Teacher educator professional learning in the third space: Implications for identity and practice. *Journal of Teacher Education, 65*(4), 315–326. doi:10.1177/0022487114533128

Wohlwend, K. E. (2009). Early adopters: Playing new literacies and pretending new technologies in print-centric classrooms. *Journal of Early Childhood Literacy, 9*(2), 117–140. doi:10.1177/1468798409105583

Wright, F., White, D., Hirst, T., & Cann, A. (2014). Visitors and residents: Mapping student attitudes to academic use of social networks. *Learning, Media and Technology, 39*(1), 126–140. doi:10.1080/17439884.2013.777077

Yin, R. K. (2009). *Case study research: Design and methods* (4th ed.). Thousand Oaks, CA: Sage.

Young, J. R. (2008, August 18). When professors create social networks for classes, some students see a 'creepy treehouse' [Web log post]. Retrieved from http://chronicle.com/blogs/wiredcampus/when-professors-create-social-networks-for-classes-some-students-see-a-creepy-treehouse/4176

Zeichner, K. M., & Gore, J. M. (1990). Teacher socialization. In W. R. Houston (Ed.), *Handbook of research on teacher education* (pp. 329–348). New York, NY: Macmillan.

Zembylas, M. (2003). Interrogating "teacher identity": Emotion, resistance, and self-formation. *Educational Theory, 53*(1), 107–127.